# Posthumanist Manifesto

## Posthumanities and Citizenship Futures

**Series Editors**
**Peggy Karpouzou and Nikoleta Zampaki**

Through the innovative interface of Posthumanities and Citizen Humanities, the Posthumanities and Citizenship Futures Series examines the changing status of subject, subjectivity, agency, humanity and citizenship, depending on the complex relationships between nature, technology, science, and culture. Given the rapid and extensive technoscientific developments, we need to conceive new species' and planetary narratives beyond our anthropocentrism. The Posthumanities and Citizenship Futures Series reflects on the possible future outcomes of humankind and defamiliarizes the mainstream narratives of humanity so it can be better understood in how it is constructed, performed, and protected. The implications of human and nonhuman life forms' co-existence within our networked world are researched in the theoretical framework of posthumanism and citizenship studies and through various fields and concepts such as literature, art, urban ecology, smart cities, Anthropocene, the future of humans, and Humanities.

Proposals are invited by crosscultural and transnational approaches, including but not limited to: environmental posthumanities, citizen humanities, literary theory, cultural studies, philosophy, animal studies, plant studies, religious studies, disability studies, narrative studies, AI and robotics, biotechnology, biopolitics, civil justice, bioethics, medical humanities, gender studies, digital humanities, art, visual studies, media studies, indigenous studies, educational and social studies, psychology and anthropology. The Posthumanities and Citizenship Futures Series seeks to foster an ongoing dialogue between academics and scholars across the globe by featuring monographs and edited collections exploring new narrations, raised by the intersection among biosphere and technosphere in a more-than-human citizenship world.

### *Recent Titles in the Series*

*Posthumanist Manifesto: A Pluralistic Approach*, by Roberto Marchesini
*Modernist Parasites: Bioethics, Dependency, and Literature, post-1900*, by Sebastian Williams
*Toxicity as a Form of Life in Contemporary India*, by Dipali Mathur

# Posthumanist Manifesto

## A Pluralistic Approach

Roberto Marchesini

LEXINGTON BOOKS
*Lanham • Boulder • New York • London*

Published by Lexington Books
An imprint of The Rowman & Littlefield Publishing Group, Inc.
4501 Forbes Boulevard, Suite 200, Lanham, Maryland 20706
www.rowman.com

86-90 Paul Street, London EC2A 4NE



British Library Cataloguing in Publication Information Available

**Library of Congress Cataloging-in-Publication Data**

Names: Marchesini, Roberto, 1959- author.
Title: Posthumanist manifesto : a pluralistic approach / Roberto Marchesini.
Description: Lanham : Lexington Books, [2024] | Series: Posthumanities and citizenship futures | Includes bibliographical references and index.
Identifiers: LCCN 2023037043 (print) | LCCN 2023037044 (ebook) | ISBN 9781666928228 (cloth) | ISBN 9781666928235 (ebook)
Subjects: LCSH: Posthumanism.
Classification: LCC B821 .M319 2024 (print) | LCC B821 (ebook) | DDC 144--dc23/eng/20231006
LC record available at https://lccn.loc.gov/2023037043
LC ebook record available at https://lccn.loc.gov/2023037044

# Acknowledgments

I would like to thank Cosetta Veronese for her dedication and commitment to the translation of this book.

# Contents

# Introduction

It is now common practice to refer to posthumanism as the philosophy of the twenty-first century. We could even say that posthumanism seems to be the favorite interpretative key to tackling the problems of the contemporary age. But what does it mean to adopt posthumanist criteria or a posthumanist focal lens to address the great questions that lie ahead of us? While the directions of thought on which it is founded are still ambiguous, posthumanist philosophy resembles a tsunami we cannot ignore that can bring together artists and intellectual energies in a constructive way. Posthumanism undoubtedly marks a paradigmatic rupture in the modern age *continuum* because of a marked disinclination to stay inside traditional perimeters and a tendency to avoid the grip of the present, on the wave of change. This perspective has elicited multiple points of view, all of which implicitly share the idea that a particular cultural paradigm—the humanist one—has definitely come to an end.

In a humanist view, a manifesto is a program, or a conceptual a priori, that aims to chart a future course, one that has yet to unfold. This is because it does not start from a hybrid and emergent understanding of the phenomenon but it is part of an essentialist and top-down view. I take a totally different approach toward the manifesto. Mine is already based on a posthumanist conception that takes the emergent process as aposteriori and bottom-up, drawing on the undergrowth of phenomena in the making. My manifesto is therefore not a program but a platform for discussion, starting from the conceptual pivots and problems that have emerged over the last thirty years. This manifesto will examine what has emerged in order to launch a phase of clarification on the problems raised by the posthumanist discussion, pointing out its differences from the humanist approach.

However, we must not fall into the trap of assuming that the posthumanist proposal leans toward an anti-humanist vision. What is in question is not so much the human as the *autarkic and autopoietic view of the human*, or what I call "ontological anthropocentrism." Posthumanism is therefore a philosophy that rejects the classic idea that the human is different from and impervious to the rest of the living world. Posthumanism adopts a relational view that considers ontology the result of the relationship between the human and the external world and subjectivity the fruit of introjected experiences. In this respect, posthumanism is a reinterpretation of the human condition in hybrid terms, namely as a process of continuous becoming, which feeds on the contributions emanating from otherness. We draw on some of the assumptions of relational anthropology in order to go over the human and understand the hybrid dimension of human predicates in an *eco-ontological* existential perspective.

By overcoming the Vitruvian metric that views the human as a measure and synthesis of the world and *dignitas* as a mark of human superiority, posthumanism draws our attention to hybridization, that fundamental principle of hospitality that makes the human a chiasmic point of contamination. With regard to this point, we can also speak of posthumanist aesthetics where the sense of the sublime regains vigor, albeit in a totally different way from Romanticism, namely on the basis of canons that can be opposite to the ideas of purity and awe. This type of aesthetics is still probing its somatic points of reference. The exhibition of the body is a testing ground for the research of new metrics. It is a projection onto the world that transforms the flesh into a phenomenological dimension, an opening toward, and a reference to, otherness.

We could speak of aesthetics as symbiosis, infection, and contamination. Hybridization is the moment of the Apollonian becoming while staying within a Dionysian flow of possibilities and transitions. Posthumanist aesthetics, therefore, is *aesthetics of mutation*: this means that although mutation is induced by the outside, it is possible because the body's constitution implicitly contemplates its own conjugation by external entities. Posthumanism is interested in the relationship between the human, technology, and the rest of the living world.

The terms of this relationship, however, have changed at the roots. The relationship is no longer one of fruition and consumption by the human. It has been reconfigured altogether, so that it can be referred to as a partnership. To change the terms of the relationship with nonhuman entities requires us to reinterpret the *concept of human nature*. This is one of this book's core points: its purpose is to conceive of an anthropology no longer based on deficiency, as in the tradition of philosophical anthropology, but on the copulative exuberance of human nature, namely on its predisposition to build predicative relationships with the outside world. *Heteronomy* is therefore opposed to autonomy and humanist autopoiesis. Its openness toward otherness is a predisposition to embrace it and realize a new form of existence. It should not be confused with the Promethean principle of the human's deficient biological equipment. Through hybridization the exuberance of human nature does not seek compensation; instead, it conjugates its contents like a verb can be conjugated in different forms and acquire different meanings. Rather than a response necessary to make up for a deficiency and achieve adaptation, *téchne* can be referred to as a dialogic and creative impulse.

Posthumanist philosophy is based on the remodeling of human predicates and the resulting new existential dimensions. It entails profound changes in all the subject's expressions, whose qualities are defined by its relationships. Hence, epistemology becomes less and less anthropocentric, ethics accommodates otherness and ontology firmly focuses on relationships and hybridization. Confident that an anthropology characterized by technomediation is irreversibly bound to decentralize the human, posthumanist philosophy brings together anyone who wishes to explore new perspectives in what is perceived as an epoch of transition. For this reason, it is hardly surprising that posthumanism displays a certain pluralism of perspectives, even though it strongly differs from both the humanist and transhumanist approach because it takes scientific advancement into greater consideration and adopts a more critical stance. Posthumanist discourse revolves around the anthropological dimension because of the effects produced on the human by the advent of new technologies and the resulting rapid ongoing changes, whose repercussions for the ecological crisis are calling for undeferrable choices.

Posthumanism reflects in phenomenological terms upon the somatic metamorphosis that the human body is undergoing. It is progressively being infiltrated by hybridizations with the technological apparatus that alter not only its performative dimension but also its predicative and teleological character. I would like to point out the most prominent differences between humanism and posthumanism. I will refer to three of them in particular: (i) the rejection of the ontological difference between the human and life phenomenology, in particular the view of animality as contralateral to the human; (ii) the overcoming of the idea of an anthropological deficiency in order to account for human predicative tension; this enables us to view copulative and creative exuberance as the trigger of anthropopoiesis; and (iii) the rejection of human ascentionality or verticalization, that attitude of self-withdrawal that leads to human loneliness, the desacralization of the entire nonhuman world, and the frenzy of transforming it into a human product. These three points are not only theoretical interpretative keys, they also help us understand some of the critical issues of the modern age. Moreover, they aspire to guide this transition by offering *a project to future societies*.

As far as the first point is concerned, humanist dualism has evidently created a separation between the human and nature that is no longer acceptable. This reflects itself in the anachronistic separation between the human and natural sciences and the resulting gap between different disciplines. Knowledge is not only constrained within increasingly cramped specializations—understandable as they may be—but it is also based on paradigmatic matrixes, which are incongruent with each other. What I would like to address is the revolution initiated by the work of Charles Darwin and its crucial philosophical implications. Combining philosophy with science, I will deliberately adopt a *hybrid approach*, because I believe we need a comprehensive vision that includes both of them. I do not think we can discuss the human if we ignore its phylogenetic traits, or if we deny that it is a conjugated dimension of animality. However, in order to reconnect ontological domains that were forcibly separated—they were defined by a game of contrasts—we need to reestablish both of them. We need to avoid both the view of the nonhuman as passive extension that has characterized humanism, and the idea of total fluidity, which explains all human predications by reference to the principle of deficiency. This does not mean overlooking the fact that the human is a particular animal, but acknowledging, firstly, that it is an animal and, secondly, that its qualities cannot be separated from its phylogenetic constitution.

As far as the second point is concerned, it is clear that the humanist ploy of the *Epimethean deficiency* has encouraged the project of ontological empowerment. This is for various reasons: (i) natural ranklessness—the human's inability to be ranked—became synonymous with freedom and self-determination; (ii) biological deficiency has enhanced the human's heroic path: notwithstanding the original disadvantage, the human could prevail on a nature that was as hostile as it was well-equipped; (iii) Epimetheus's forgetfulness made it possible for us to become the children of Prometheus, endowed with wit and foresight rather than stunned by instincts; (iv) the genealogical difference between humans and animals justified their ontological difference; and (v) the adaptive deficit justified the advent of *téchne* as compensation and a means to redress our functional incapacity. To demolish this big castle obviously means questioning the very roots of modernity, whose foundations were inspired by Pico della Mirandola. To question the principle of deficiency, however, does not mean affirming autonomy, because, on the contrary, it was on the very idea of deficiency that humanism built its autarkic and disjunctive framework. To posit that the human is rooted in *life phenomenology* means bringing human creative tension back into the exuberance of *bios*, something which, in our species, is expressed through technopoietic and symbolic projections.

From this, we infer a third point about the difference between a vertical view of the *becoming-human*, typified by humanism, and a horizontal view, based on our accommodation and projection into the biosphere, epitomized by posthumanism. The vertical view creates the premises for a divorce that pushes the human toward absolute extraneity from life phenomenology. Transhumanism turns out to be the legitimate heir of that expression of humanism whose peak is represented by the emancipative rationalism of the Enlightenment. Significantly, transhumanism postulates that the human's future will involve a postorganic dimension, digital transcendence, and a transformation of the mind into a packet of information transferrable via mind-downloading. Verticalization, however, does not simply imply distance. It also means looking down on the nonhuman, stressing its extension-panoramic character and considering it a cheap resource available for human use. Posthumanism questions the human's dominion over the world and believes this view to have been the fatal error of humanism, causing the critical issues to which modernity is heir. All human activities are two-sided: on the one hand, they increase our degree of intervention in the world, while on the other, they increase our need for the world. Offering a univocal interpretation of technology's potential reveals that we cannot fully understand it.

This is the reason why we need to distinguish between: (i) the transhumanist vision that considers *téchne* a principle of human empowerment, hence a tool that furthers human verticalization and may fulfill the humanist dream of emancipation from nature; and (ii) the posthumanist vision, for which, on the contrary, relationship and the preservation (actually, the enhancement) of the subject's qualities and relational opportunities are paramount. This network of relationships moves horizontally inside a view of *ecological conviviality*. The great ecological crisis of our time—assessed, at best, in terms of global warming, while being, in fact, a much broader and complex problem—is at the heart of the posthumanist proposal which, in some respects, can be called ecological humanism. Ecology hybridizes with philosophy and reveals how some of its propositions—the relational emergence of predicates, for example—may have important heuristic implications for ontology. Technomediation is also interpreted in ecosystemic terms as a niche that transforms the morphogenetic parabola of a species, rather than a shell that enshrines an enhanced individual. Posthumanism does not conceal the risks of a techno-capitalist version of *téchne*. For posthumanism, we need a critical attitude that avoids forms of techno-enthusiasm—and, even worse, of the soteriological projection of technology, as in transhumanism—as well as of technophobia and neo-Luddism.

The purpose of this book is also to prevent confusing posthumanism with a hyperhumanistic fascination for the horizons of human ingenuity, which happens quite frequently today. Posthumanism is not about being intoxicated by the magnificent progressive destiny of humanity. It is not yet another expression of Prometheanism, a fetishistic *pathos* that makes the human a Pygmalion in love with his artifacts. It is exactly the opposite. Posthumanism is a critical reflection upon all this, a tragic staging of these passions, of their value as emotions and desires. To reflect upon *téchne* means to probe the depths of the human dimension and try to understand its reasons and risks. As a relational philosophy, posthumanism focuses on the crisis of the biosphere and our difficulty of reasoning in ecological terms—which explains why our discussions are limited to topics such as eco-compatibility or the *green economy*. To think of *téchne* in critical terms means, for example, to disclose the paradox of being witness to a technological evolution toward self-programming and emancipation from human control, while technical philosophy still views the instrument as something we handle and in terms of compensatory ergonomics. On the one side, there are artificial intelligences that develop without needing a programmer to define their functions and on the other side, we continue to believe that these tools are under human control. Posthumanism clearly reveals that a new epoch is coming on the horizon. This epoch requires us to change our terms of reference and our way of interpreting problems. Apart from a generic *pars destruens* comparing the posthumanist to the traditional approach, I believe we also need to take into consideration the core points pursued by posthumanism in order to realize this paradigmatic change.

The points of humanism criticized by posthumanists include: the tendency toward oppositional dualism; the ascensional view as a means to achieve human *dignitas*; the myth of purity and incompleteness; the self-referential and autopoietic view of anthropopoiesis; the idea of the subject's autonomy or the view of nature as mere extension (*res extensa*); the ergonomic and instrumental interpretation of *téchne*; the contralateral view of animality (human *vs.* animal); the idea that the human, symbolized by the Vitruvian Man, is a measure and subsumption of the world; the diffusion of *anthropoplasty*, namely the idea that nature is something to be converted, Pygmalion-like, into a human product; self-withdrawal as a means of delving into the sense of human existence; and the notion of *Thrownness,* the condition of the castaway in the universe. It is not sufficient to speak of a human metamorphosis in order to lay new philosophical foundations. The risk would be to remain at a purely descriptive level, implicitly and misleadingly preserving the system of meaning and the hermeneutics of the humanist tradition.

All of these needs motivated me to formulate the *Thirty Theses*. They are proposals for reflections upon some of the major issues of posthumanism. Their goal is not so much to build an exhaustive platform as to stimulate dialogue and research through ideas that move in a specific direction and have an internal cohesion. I believe that a philosophy that really claims to be posthumanist cannot help addressing the great ontological, epistemological, and ethical questions that inevitably arise when we modify the order and scale of the factors at stake. Otherwise, we are only talking science fiction or, at best, futurology, concealing the smallness of our reflection behind the propaganda lights of a great technological amusement park. In doing so, we skip uncomfortable questions, such as the issue of the predicative meaning of the human and the consistency of *Being-there* in a digital flow without space or time. This is certainly not a philosophy, because there is no questioning.

Even less is it a paradigm, because it has no foundations. Posthumanism, on the contrary, poses an unavoidable question: what does it mean to be human from a perspective that distances itself from the foundation *iuxta propria principia*? We could even ignore the technoscientific exaggerations and focus only on the relational significance that posthumanism considers to be the central topic of ontological analysis, because this is exactly what we should be talking about.

Over the last few decades, a great debate has developed around the *concept of the posthuman*. But little clarity has perhaps been made about the meaning of the term. It refers to an ongoing metamorphosis—the act of going beyond a domain without yet having a sharp picture of the emerging perspective. The posthuman is often confused with an entity that is no longer human, or with an overcoming of the human or, once more, with empowerment: a futuristic vision of this transhumization had already been predicted by Julian Huxley. Yet I do not believe that this rather banal interpretation, halfway between the American superhero and the Japanese manga, is the most appropriate one to discuss posthumanist philosophy, a philosophy that aims to overcome the humanist paradigm through a new concept of the human. The posthuman is a *condition of the human* illuminated by the awareness of its own hybrid dimension. According to posthumanism, the hybrid is a presence that emerges from a conversion, not from a futuristic projection.

The posthuman addresses the question of openness, the intentionality of life phenomenology and the sense of ontological symbiosis. "Overcoming" does not refer to biology or performance but to the constitution of the predicate, starting from a *conjugative conception of the entity*—the entity conjugates like a verb. If the humanist vision was founded on autarky, self-determination, and autonomy, the posthumanist one reverts to an eco-ontological view of the human, where each predicate emerges through a relationship. The terms with which to discuss in talking about posthumanism are these, rather than biological or performative ones.

This is the purpose of the *Thirty Theses*: to reflect upon how humanism has bequeathed us a very distinct profile of the human, and to consider whether, starting from different assumptions and with equally different purposes, it is possible to reformulate it. While humanism considered the human an entity tending toward disjunction and verticalization, we wonder whether it would not be more productive and coherent to take into account the human's copulative tendencies along with its hospitable and caring qualities. This is just one example that helps clarify that hybridization does not necessarily refer to the enhancement of a subject's egoic and narcissistic tendencies. It can be interpreted as an openness to otherness and the awareness of *Being-with-another*, something that characterizes an individual's entire ontological parable. This book, therefore, would like to offer a new interpretation that has arisen among

Italian authors in particular. According to this reading, the posthuman is neither a new biological or postorganic entity that will take over the human, nor the goal of a humanity that has hybridized with technology or been surpassed by it. The posthuman that emerges from this book is a new existential dimension that acknowledges its hybrid condition and openness to the world.

We can argue that the term-concept that comes into being in the forthcoming pages is neither a postorganic entity nor an artificial intelligence nor a cyborg infiltrated by technology. What makes it posthuman is not its performative profile but the acquired awareness of its connected and hybrid nature. Its ontological basis goes beyond its phylogenesis and gains new dimensions: in the psychagogy of altered states of consciousness; in the shamanic art of projecting itself into animal otherness; in the use of plants with psychotropic effects; in the cultural mimesis of nonhuman strategies and choreographies; in the fashioning of the chipped stone and in the art of fire or the use of clothes. In this sense, the human has always *been over the human*: we have followed the fascination of ecstatic centrifugation or possession, we have learned from the world and decentralized. We have been hybrid entities from the start, we have not built our existential qualities and dimensions autonomously, but by means of conjugations based on otherness. *We have always been posthuman, but it is now time to gain awareness of it.* Posthumanist philosophy cannot remain a perspective: we must build a retrospective. In other words, we need to interpret the history of humanity through this paradigm and view the human as a copulative threshold whose predicates have always emerged through relationships.

Posthumanism is thus our need to grow empathy, which may diminish the solipsistic and narcissistic risk of anthropocentrism: this is imposed on us by the ecological crisis and the development of intelligent technologies, which will otherwise overwhelm us. Just as the empathic individual can understand the others' concerns, so an *empathic humanity* will need to rediscover the propensity of feeling part of the whole, in a *convivium* with the living universe. On the other hand, a truly new posthuman consciousness cannot limit itself to making moralistic prescriptions about our need to respect the biosphere. We need to understand that no entity—not even the human—can exist within a logic of autonomy. Besides making us see things in a different light, this new awareness stimulates reflections that need to acquire critical consistency. Its inclination for heteronomy has given the human enormous possibilities, which have been absorbed into culture. This does not mean, however, that the progress of an increasingly infiltrative technology that controls our choices might not become a dangerous vulnerability. The posthumanist theses here presented do not fall into the trap of rejecting technology. Yet by taking a critical attitude, they also eschew overenthusiasm or a salvific view of technology.

We cannot speak of the posthuman without questioning the traditional philosophical guidelines that depict the human as an entity who gains self-awareness through distance, suprematism, the trivialization of otherness and by refusing to acknowledge the debts it has contracted with the world. The term "posthuman" does not indicate an overcoming of the human, but of the humanist framework. It is also true, however, that a phenomenological divergence inaugurated by posthumanism has developed around the concept of human. Human is no longer the emanative expression of *Homo sapiens*, but the outcome of the hybridizing events produced by *Homo sapiens*. Watching birds flying it is not like observing flying machines: the human identifies with the birds' existential dimension. In other words, *the human becomes a bird*, it dreams of flying and experiencing the vertigo and exhilaration of flying. This is what we mean by hybridization: when, little by little, the use of the chipped stone transforms the tool into an organ of the body. Hybridization is not only about gaining a performative potential, but about transforming one's existential dimension. The hermeneutical switch that posthumanism produces originates here: human predicates and the anthropocentric dimension no longer overlap. According to the *Thirty Theses*, the human finds accomplishment in anthropo-decentralization.

Without a new philosophical perspective, the posthuman would remain an empty vessel, a blank sheet to be filled in with various technical illusions, more akin to sci-fi imagery than ontological reflection. It does not make much sense to talk about cyborgs if we are to adhere to an essentialist view of the entity's predicates, and if we have to derive its contents by studying it from within. In doing so, we lapse into idle or aporetic questions: such as to wonder whether it is technology that will use us, or vice versa; or whether, in a condition of mind-downloading, we can still call ourselves human. What is paradoxical is claiming to present new teleological theses while holding onto traditional ontological guidelines. According to the latter, the human stands at the center of the universe, is the measure of the world, can subsume all other entities, is above nature and opposed to animality, is self-determined because it is biologically deficient, is Promethean and therefore a craftsman, is a craftsman but also a trickster, is surrounded by technical equipment but pure in its dimension, and possesses a dignity that needs to convert nature into a product in order to feel accomplished.

In order to speak of posthuman, it is necessary to challenge these assumptions. We need to do so, however, not because threatening or salvific artificial intelligences are urging us to do so, nor because of our serious ecological crisis. We need to do so because it is the premise by which to understand our human condition: its prerogatives, weaknesses, conflicts and ultimately the reasons for the problems it is going through. To do this, we have to avoid nihilistic and anti-humanist temptations as well as the enthusiasm and triumphalism that still nurture solipsistic inclinations and anthropocentric arrogance. To prevent both attitudes, we need a critical perspective that can trust the human to be able to build a new alliance with the biosphere and view technology as a mediator. But it must be a *critical trust*, not an unconditional or self-justifying trust. To acknowledge one's own hybrid and connected condition means to rediscover a sense of one's limits. This realization should not come from the outside, but be introjected, through the achieved maturity of reason and feelings and a deeper self-awareness.

The human has created its own mythological image through many stories that have flattered its anthropocentric vanity and encouraged its arrogant attitude toward the world. This has certainly not contributed to harmonious human development. It has actually laid the foundations for a disharmony directed toward both the outside and the inside, a dystonia that led humanity to drift away from all kinds of relationships, not only with the biosphere or the technosphere, but also with itself. The idea of natural human indeterminacy has prevented us from coming to terms with our inclinations and knowing how to redress them. We must start by realizing that our species is the *outcome of the phylogenetic process* that molded our innate tendencies—some of which are a source of criticalities. Yet another problem arises here: how can we analyze these inclinations? We are used to thinking of them as the yardstick of biological mechanisms, determinants that can be contained or managed by reason. We should instead view them as the interpretative key to life phenomenology, references that come into the world in search of a conjugation. In the former case, we read our inclinations as factors that hinder the emergence of the human. In the latter, they become flywheels of anthropopoiesis. This is the stand we take in this book.

I believe that there is a profound fissure between the humanist view of nature as pure mechanism and the posthumanist one according to which nature is a diachronic entity that cannot be explained outside a process of becoming in time. *Having-been-there-before* is a core concept related to this point. It expresses the diachronic condition of the subject, who is no longer confined within the boundaries of its own individual and species-specific experiences. This is a reading that reinforces the Heideggerian concept of *Dasein*, and the correlation between the subject and the world, which we find in the panorama of phenomenology. But there are significant differences, because my concept distances itself from the idea of *Thrownness* The image that best depicts the condition of finding oneself again in the diachronic flow of being is that of a smiling baby who first encounters its mother's face. The pre-subjective dawn does not discover the world but recognizes it when it appears, because the world has already shaped the emerging subject via its phylogenetic legacy.

*Having-been-there-before* encompasses the flow of everything preceding the individual, who comes into the world to gain subjectivity by giving form to this magma. Unexperienced memories that are not even totally human are the web of an individual's correlation to the world; they already speak of the world. This is why our first experience of life is one of joy and wonder for a *déjà vu* that lingers on the horizon. We rediscover the world thanks to our internal constitution. Therefore, the real cannot be ascribed to a logocentric hypothesis but to an ontological correlation. To refer to pre-experiential inclinations, therefore, does not mean to fall into a deterministic view of nature, but to discover the intentional meaning of life phenomenology, a meaning that contains many references to the world.

The tension of the *Having-been-there-before* originates from a principle of desire that can never be ascribed to a lack or to the object-result of an action. It refers to a creative and copulative action. To desire is a verbal predicate without a direct object. Human nature is the driving force of actions whose results are always unpredictable because the meaning of the action is not in the object, but in the relationship. To speak of human nature, therefore, does not mean lapsing into reductionism or biological determinism. It means understanding what the intentional bases that allow our opening toward the world are. These inclinations do not oppose our subjective expressions: human nature is copulative, open, and in love with the world. Our legacy is the most important flywheel of every expression of singularity. There is a striking difference from the humanist approach, which separates nature from culture and sets them in opposition to each other. The humanist dualistic structure offers a model for further oppositional dichotomies. On the other hand, to reflect upon human nature avoiding the idea of human autonomy leads us to: (i) reinforce the critical trust granted to the human; and (ii) analyze the possibilities that these inclinations offer for a good life, not only a fulfilling life, but one that also produces values that go beyond us. Posthumanism is therefore a form of eco-ontology that views a relationship as a predication.

Hence the sense of *post*humanism: its prefix indicates a going-beyond not in order to wipe the slate clean and erase all it has produced, but to insert it into a more solid and better-grounded frame. This philosophy urges us to exit the childhood of humanism, which is still self-centered and leads us to believe that everything is at our disposal and all ideas are the fruit of our ingenuity. Since every dimension needs an infantile and apparently self-centered phase of development, the human has probably also needed a period of joyful expansion in the world, a solipsistically dreamy age like the Cartesian cogito. The humanist of today, however, is a spoiled and whimsical teenager who cannot understand what future lies ahead of him in relation to technological development and the ecological crisis. He is a muscular hyperactive guy, bulky and destructive. But he also whines about having been forgotten by Mother Nature. His cyclopean ego perceives threats in everything he cannot control, and he is led to believe that everything is permissible. Misled by his own successes, he thinks that whatever he does is a masterpiece that comes out of him and him alone, without realizing that his creations are the depletion of previous resources.

The task of posthumanism, therefore, is to institute a true *seminary for the youth* in order to help the human achieve a level of maturity based on its awareness of *Being-in-relation*. Respect for otherness should become an integral part of human existence, without the need of imposed rules. It is also misleading to think that the problem is the human as such as well as to speak in generic terms of *anthropocene*. We forget that there are different levels of responsibility for the world's current conditions and that not all humans are equally responsible. The question should be posed in cultural and economic terms and the revision can only be paradigmatic and should not be based on an insane project of self-extinction. The purpose of the *Thirty Theses of the Posthumanist Manifesto* is to hypothesize new forms of citizenship and a society that can call itself post-Neolithic. This society ought to rethink the very guidelines of the Neolithic Age, such as patriarchy, the logic behind war, capitalism, agro-livestock, the politics of fire (the use of fire to control the territory), and the subjugation of nature. What it all comes to in the end is a *new sentimental education*, conceived as the comprehensive, rather than separated, contributions of reason and emotions, nature and culture, humans and the biosphere.

Today, it is essential to overcome anthropocentrism, namely the perspective that separates the human from the biosphere by treating the former as an autonomous entity that can exploit all that is around it and convert it into a resource. Anthropocentrism is a model that fails to understand the anastomosis of the human with the rest of the living and forgets its symbiontic nature: other forms of life not only populate, but also make up the human organism. Instead, we keep on thinking of the biosphere as a home in terms of inhabitability rather than common metabolism. This inevitably leads to visions and projects that are misleading even when we naively formulate environmental perspectives. As a philosophy of relationship that is called upon to express a hermeneutic of fusion between body and technology from an angle that is completely different from ergonomics, posthumanism changes how we interpret our relationship with the biosphere. We do not inhabit the biosphere: we are integrated in it. Posthumanism, therefore, tries to bring ontology back to an *eco-logical* level by systemically rethinking philosophy: the predicate expresses a relationship and not an essence. We are, because we are connected.

We cannot speak of posthumanism without reinterpreting epistemologically our relationship with the real in non-anthropocentric terms. This means that we need to step away from the universal view of the human (the human as a measure of the world), from the anthropomorphic view of the real (the world as a mirror of ourselves), but also from the idealistic view of the real (the world as constructed by us). To reason in relational terms means to erase the subject-object dualism, because every single perception is always both an integration and a referral. We could argue that the object is never totally alien to the subject, even when it is unknown, because every form of knowledge requires a shared element. This is due to the diachronic nature of the subject, who is defined by its experience, namely it has been shaped by the world. Arguing that a prioris are phylogenetic a posterioris casts some light through the looking glass. Yet we do not always remember the trans-specific character of this configuration. It is not only the subject who observes the world, but the world that mirrors itself in the subject. The correlation between subject and world leads us to distance ourselves from both Cartesian doubt and Kantian criticism not because the real is just as we perceive it, but because it is consistent with our systems of knowledge. Perception is the consequence of the type of diachronic relationship that that particular entity institutes within a range of possible conformations.

In this sense, the human cannot be defined in univocal and static terms, either, because it actually expresses the relationship that every subject establishes with otherness. This enhances the difference between the humanist and posthumanist positions: the former is based on the logic of *deficiency*, whereby the relationship turns into either compensation or opposition; the latter is based on the logic of *redundancy*, for which the relationship is a conjugation of the entity's inherent and overflowing potential. This means that, although we start from a state of virtuality, we are not confronted with an amorphous condition or one only shaped by the outside. We are confronted with a range of possibilities whose qualities are produced through a dialogue with the outside. This eco-logical view of ontology, where even what is innate is the fruit of previous relationships—the *Having-been-there-before*—causes an important existential shift: from *Being-in-the-world* to *Being-through-the-world* or *for-the-world*.

Posthumanism changes how we consider the entity's manifestation. The entity does not disclose contents that are exclusive—essential—to it, but expresses the encounter with otherness in the here-and-now (conjugations). Inherences also need to be seen as the disclosure of previous hybridizations, which can be either phylogenetic or ontogenetic. Therefore, we must create a new ontology that may overcome an antinomic and dualistic conception—relationship as *polemos*—but also one that does not level differences through forms of monism or shapelessness. Relationships produce predicates that emerge through the encounter with otherness but are regulated by dialogical inherences. The inherences that interrogate the world and open up to its presences, that host its alterities or project themselves into them, are nothing but the fruit of all the relationships that the entity has introjected before coming into the world. In this sense, posthumanism is a philosophy that reconciles the human with the world. It overcomes a pessimistic view of existence that has been responsible for countless forms of alienation and equally misguided attempts to remedy it through solipsism and technological regression. *Téchne* should not be regarded as armor, but as a means with which to enlarge our windows onto the world and increase our fragility, which is also an antidote to arrogance.

What characterizes my posthumanism is an epistemologically hybrid approach, somewhere between scientific and humanistic approaches, between biology and philosophy, that reconciles the Darwinian and ethological vision of a fully animal human with the Nietzschean one of an existential dimension that is constantly being transcended. The plasticity that I confer on the human condition does not, for me, consist of a lack of nature—and consequently cannot result in total fluidity—but lies, on the contrary, in redundancy. For me, posthumanism means first and foremost rejecting the disjunctive conception, which belongs to humanism and views the human as separate from nature and animality. Posthumanism is a critique of the humanistic assumptions that make nature deterministic and the human detached from it. Human freedom lies in nature, not in the lack of nature. It is nature that creates the virtual condition that makes the human a continuous becoming. For me, posthumanism means reconsidering nature as a creative rather than a mechanistic and repetitive entity. Its natural characteristics make the human a copulative entity who can be transformed through the contribution of otherness. Hybridity therefore lies in the copulative tendency and the contribution that animals and plants have made to the emergence of humans.

These *Thirty Theses* are my attempt to put on the table of philosophical discussion arguments that I deem fundamental to the posthumanist debate. They also aim to encourage further research projects. I would like to offer suggestions and promote a *pars costruens*, which, in my modest opinion, has been missing. I am offering some suggestions and ideas that require further development and critical analysis but I believe this manifesto to be a good starting point. Posthumanist philosophy obviously cannot, and should not, limit itself to a critique of the traditional humanist vision. It has an important task to fulfill: to draft a proposal for the future that begins with the key word "reconciliation." I think it is greatly needed in many aspects of our lives.

## *Chapter 1*

# On the Human Condition

This chapter analyzes the debate about the human condition, which stands at the forefront of posthumanist philosophy. Posthumanism challenges the essentialist view of the human that has developed since the fifteenth century. For the early humanists the human was like a radiant entity, both the center of the universe and the measure of all things (Kemp 1990). For posthumanism instead the human is hybridized with reality: rather than being the center of the world, it merges with the world. A new model of philosophical anthropology thus surges forth. The theses addressed in this chapter suggest a different way of considering human nature: the human is a redundant, rather than deficient entity, and, at the same time, a nondeterministic entity. Its phylogenetic richness is precisely what gives the human ample room for cultural conjugation. The chapter also deals with the theme of animality. Animality is not the dark counterpart of the human any longer, but a shared dimension, which each species interprets in distinctive ways.

### FIRST THESIS

The human does not express its predicates in a pure form. It cannot be considered an autarkic entity nor be evaluated in essentialist terms because its manifestation is always based on hybridization with otherness.

We can infer the hybrid condition of the human from the pervasiveness and complexity of its cultural and technological apparatus. Posthumanist reflection goes deeper than merely observing how culture has affected the human dimension (Ferrando 2019). Posthumanism criticizes (i) the autarky of the technopoietic and cultural processes: these are considered relational outcomes rather than the solipsistic results of human ingenuity; (ii) the idea that the cultural apparatus provides compensation and empowerment, influencing all the expressions of the ontological condition—the epistemological, aesthetic, ethical, and existential ones; and (iii) the idea that the apparatus is external or disembodied, that culture is like a garment. For posthumanism the different domains are not separated but mutual; the apparatus modifies the body, and the body incorporates the apparatus.

It is impossible to sift the purity of the human, to extract its uncontaminated genetic content, the essence of the species. This is because each of its characteristics (even hereditary ones) reveals the influence of hybridizers on phylogenetic selection (Gilbert and Epel 2015). Even more so during the ontogenetic process, when multiple influencing factors—social, cultural and economic—cause segregations and differentiations, showing the extent to which predicates are dependent on their counterparts. The humanist vision has accustomed us to a radiant image of the human, who, like a sun, is the pivot of a world orbiting around it, or the peak of a pyramid, its predicates shining like starlight (Manetti 1975). This view not only led to a hierarchization of the human, but also condoned an autarkic view of the predicate, of its development and expression. This idea explains why the human was considered to have been founded *iuxta propria principia*, and its predicates could be derived from its phylogenetic qualities.

If we consider the cultural product no longer as emanation but as hybridization, both the autarkic conception of the apparatus and the principle of purity collapse. The predicates of the human are no longer rigorously "its own"; they are the result of a cultural symbiosis which is the fruit of its relationship with otherness. Inevitably, the images by which we define the relationship between the human and the world, and which help us identify human singularity change, and we withdraw from Vitruvian anthropocentrism. In order to illustrate this change of perspective, we cannot just retouch Leonardo's image and add some prostheses to the joints of its body (Haraway 2016). Firstly, if we are to define this change of perspective, it no longer makes sense to use the metaphor of the human body standing in the center of a circle and stretching out its limbs. This logo-concept cannot withstand a logic that seeks to overcome the heliocentric image of the human where the human and universal overlap.

Secondly, we evidently cannot and must not speak of verticality, center and orbit. In the course of anthropopoiesis, in its becoming cultural, the human has not moved vertically (Sloterdijk 2014) but quite the opposite. In fact, anthropopoiesis proceeds horizontally, as a contaminating and not a purifying process, it stretches into the world by merging with otherness, thus increasing its dependence on the outside world rather than becoming emancipated from it. Anthropopoiesis crosses thresholds, it neither builds fences nor marks out boundaries. This attitude strikingly clashes with the image, albeit retouched, of the Vitruvian Man. The cultural apparatus increases its ontological dependence on external sources as referents: the human's hybridization with birds, for example, has created a constellation of ontopoietic meanings, which connect us with the ornithological otherness. In other words, whenever I allude to the existential dimension of levitation, flying or ascent (the wings of angels), I inevitably refer to birds.

Anthropopoiesis is therefore a decentralizing phenomenon. It is the opposite of the centripetal direction of anthropocentrism. It is centrifugal and departs from the past and its legacy. Anthropopoiesis dislocates the human from its center not because of a lack of content (de Benoist 2010), but because of the centrifugal force of its content. In other words, the human is directed toward otherness and led to hybridize with it. Through this encounter, nonhuman content is grafted on the human (Marchesini 2017). Arguing that culture is neither compensation nor emanation means that we cannot understand culture if we limit ourselves to studying the human. This is the greatest fault of the humanist tradition: to think that human predicates can be understood by starting with the human.

Centrifugal force is openness to the world, it is *mimesis* interpreted as a welcoming factor and it is human art conceived of not as a *reification of* but as a *participation in* the phenomenon. *Mimesis* is therefore not a detached representation but an expression of hospitality: it means both being possessed by, and projecting oneself into, otherness (Eliade 1964). This thesis offers a warning against essentialism in human ontology and against the human's existential solitude that represents the disease of humanism. To prevent mortifying its existence in withdrawal and isolation, in the tautology of *Being-in-life* or *Being-for-death*, the human decentralizes itself and devotes itself to otherness by taking care of it. Care is the act of greatest existential significance because it is consistent with the human's decentrative ontology, its *Being-for-the-other*. *Thaumazein* is the response to the centrifugal force: it is the ecstatic moment that helps the subject escape from solitude and partake of the world's spectacle.

Human autarky is the greatest error of humanism. It also caused most of the suffering of the modern age, when the human felt lost in the total solitude of the infinite cosmos. The grief of the castaway (Heidegger 1962) and the resulting need to find a radiant other place where to project itself are due to the human's lack of a founding core. And yet we risk falling into the essentialist trap again if we consider this core an essence from which contents can be extracted. Instead, it is a centrifugal impulse toward the world, an epimeletic opening toward otherness (from the Greek *epimeleomai* = I take care of) (Mortari 2015). Our copulative tendency is not a wealth of predicates searching for a field of expression, nor a deficiency that needs mending by external help; it is rather a catalyst eliciting processes of hybridization that produce new contents.

*Mimesis* as participation in the world means that by connecting with otherness we can experience an existential transformation, an epiphany or a predicative transfiguration. The subject never perceives itself as such, but through a process of metamorphosis, by mirroring itself in otherness. Thereby, epiphanic otherness does not appear as *other-than-oneself* but as *other-in-oneself*, an eccentric condition that, by slowly being incorporated as inherent to the self, progressively loses its value as "other" (Lévinas 2001). Contrary to humanism, which is based on an essentialist ontology, posthumanism expresses a relational ontology founded on hybridity instead of exemption. This has consequences for principles such as purity, ascent, emancipation, perfection, autopoiesis, and self-determination (Remotti 1996). Anthropopoiesis is not a way for the human to disconnect from the world, but to absorb and project itself into the world, thus increasing its dependence on it.

Although it causes fragility, dependence is not toxic (Pulcini 2009). Anthropopoiesis is an event of anastomosis with otherness, which strengthens the human need for dialogue. Humanism celebrates the human by following the canons of the Vitruvian Man—as the measure of the world, the subsumption of reality, and the principle of plasticity (Alberti 1485). Posthumanism, on the other hand, emphasizes not only the condition of human fragility but also its value. This fragility can only find comfort in taking care of the other. In humanism, anthropoplasty is a way of reducing the nonhuman to a human scale in either ontologic or instrumental terms. In posthumanism the opposite process occurs: it seeks to understand the human in its manifold relationships with the nonhuman. Human power is not that it can dominate the world, but that, through the principle of hospitality and the awareness of its debt toward otherness, it can take care of the world.

Epiphany is a fundamental moment of the hybridizing process. The latter is never conducted entirely by the human, let alone directed by goal-oriented and predefined guidelines. The human is neither the helmsman nor the director of this process, but simply a cofactor; and it is in hybridization that new possibilities open up a posteriori. Hybridization is a dialogue on which we only have limited control, because its guidelines emerge in the course of the interaction; every cultural emergence is like an open construction site. Refusing emanative essentialism means that we refuse to consider the cultural product only as the fruit of human ingenuity, and that we reject an anthropology which, by viewing otherness just in terms of "good-for" (Lévi-Strauss 1963), celebrates the human as the sole protagonist of the process.

The copulative nature of humans directed our curiosity first of all toward natural otherness, mainly animal otherness. Posthumanism is the spokesman for a shamanic transformation typical of the cultural act: *becoming-the-animal-other*, namely, building a hybrid identity through the relationship with the heterospecific, turns out to be one of the most important, if not the primary, source of anthropopoiesis. This is an epiphanic process: encountering animal otherness enables the human to see and recognize itself, not narcissistically, though, but in theriomorphic predicates (Marchesini 2018). Animal epiphany is therefore a revelation and inspiration that guides a metamorphic journey climaxing in anthropopoiesis. The encounter with animal otherness offers the possibility of experiencing new existential dimensions: the predicative space is not already given but *in fieri*. In other words, it is not about acquiring the virtues of the hybridizing animal or structuring a simple zoomorphy, but about creating a hybrid space that transcends the characteristics of the two hybridizing partners.

Considering anthropopoiesis a hybrid process means redirecting reason to its primary function, namely realizing what the consequences of metamorphosis are and preparing the institutions accordingly. Anthropopoiesis is regulated by wonder, not by rational planning; by the desiring copula, not by the definition of a goal. It belongs to the domain of dreams, feelings, the unconscious and, sometimes, altered states of consciousness (Huxley 2004). The value of psychotropics, hence of the neurobiological hybridization with particular plants, becomes clear. Culture is also the result of alliances with the plant world, which enabled humans to enhance their epiphanic and transformative capacities.

## SECOND THESIS

The posthuman condition is not an ontological future, an overcoming of the human, because the human has always been hybrid; the posthuman condition is about awareness, gaining consciousness of this non-self-sufficiency.

Being aware of the hybrid condition of the human requires a radical ontological change. In the humanist tradition, the cultural apparatus was viewed as a garment, the human as impervious to the world's contributions, and its purity as an essence that could be extracted any time. This view was ensconced in the essentialist idea that the anthropopoietic dimension was the fruit of either compensation or emanation—both autarkic ontopoietic processes. The contribution offered by nonhuman entities was undeniable, but they were instrumentalized as "good-for" (Malinowski 1973), namely as objects rather than dialogical entities capable of modifying the human perspective. They were denied an active role and considered inert tools available for human use and the pursuit of human goals.

With reference to human autarky, considering human ontology the fruit of hybridization with the nonhuman introduces two major differences: (i) firstly, the ontopoietic event or the emergence of the ontological dimension turns out to be the co-factorial result of an active role played by the nonhuman; and (ii) secondly, our way of looking at the cultural apparatus changes: no longer considered a garment or a form of empowerment, it becomes ontological dependence on otherness. We are facing a new awareness based on the realization that our ontological dimension is not self-sufficient. As a hybrid, the human is no longer the measure of the world, and not even the measure of itself. The posthuman condition marks the beginning a new epoch based on the awareness of eco-ontological participation.

The difference between the posthumanist and the transhumanist vision has been analyzed from different points of view, starting from the hyperhumanist character of transhumanism (Marchesini 2002): the latter does not seek to overcome the centrality of the human, but to empower the human through technology. While humanism (Pico della Mirandola 1942) advocated a spiritual ascension, the transhumanists of our time aspire to a technomediated verticalization; technology not only makes us ontologically superior but it also has a saving power (Savulescu and Bostrom 2009). Transhumanism views the posthuman as an upcoming condition, a horizon to reach or the necessary evolution of our species. For posthumanism instead the posthuman condition is nothing but the existential dimension of the hybrid (Kurzweil 2005).

In other words, if we consider the hybrid the existential dimension of the posthuman, we can argue that we have always been posthuman (Sorgner 2021). What does this mean? We are becoming aware of how inconsistent the essentialist reading is while becoming conscious of our posthuman condition. However, it is a process that needs commitment and effort, rather than simple observation. We must relinquish our old humanist view of ontology and the idea that there is something such as a "human essence." This is the only way to inaugurate a new epoch that will modify our key philosophical points of reference. The posthuman era has not reached full maturity yet and the risk of remaining within a hyperhumanist framework is still high. Possible mistakes include a view of technology as empowerment, of culture as emancipation, of human dignity as ascentionality, of human presence as disconnected from otherness, of ontology as anthropocentrism and of anthropopoiesis as the product of Promethean ingenuity. All this we find in transhumanism.

I would argue that the technoscientific development promoted by physics and biology in the twentieth century has compounded the anthropocentric temptation of the early fifteenth-century humanists. Even though a pessimistic vision of the future emerged during the short century (Hobsbawm 1994), human arrogance moved in lockstep. The development of cutting-edge technology in the fields of information technology and biosynthesis has not encouraged the growth of a posthumanist consciousness in the full sense of the term. Rather, it has produced a degree of human fluidity (Bauman 2000) and prompted a reflection upon hybridization, which led to contradictory results, as demonstrated by the opposition between posthumanism and transhumanism (Marchesini 2002). The current panorama is much more multifaceted than the one we could observe at the turn of the century, though. The ecological crisis has helped call human-assumed self-sufficiency into question, imparting profound changes to the ecologism of the past century that was still characterized by prescientific styles and nostalgic and retrospective conservatism.

Accordingly, the posthuman era is a philosophical project encompassing all spheres of human existence. It has concrete repercussions on the social, economic and political levels. Yet, it requires an effort rather than the mere acknowledgment of our hybrid condition. Because the risk of reframing this philosophy within the guidelines of humanism is high. Posthumanism actually wants a future conscious of what the implications of a relational ontology and of ontological non-self-sufficiency are. But what does it actually mean, to overcome the autarkic conception or to consider otherness as ontological symbionts (Haraway 2016) instead of mere tools for one's own ends?

In the footsteps of Deleuze, the word "posthuman" could easily be taken to refer to the process of going beyond the species without questioning autarky. However, by reasoning in these terms, we would do nothing but resume the humanist text and patch it up with our new scientific awareness. When I speak of a posthuman era, I do not at all mean transcending the species. We cannot go beyond the species; we can only add new conjugations to our phylogenetic heritage. What I mean to do with the posthumanist manifesto is to surpass the idea that the institution, to quote Gilles Deleuze again (Deleuze 1955), or the cultural apparatus is the fruit of the human's solipsistic imagination, the expression of a human essence. This shift marks a point of no return. We cannot relapse into a humanist framework. This shift paves the way to the posthuman era by abandoning autarkic Prometheanism. We have always been aware of the contribution of external entities. Yet we would rather believe them to be passive tools available for the pursuit of our independently planned goals, rather than ontological referents.

In a logic of self-sufficiency, we believed that studying exclusively our phylogenetic qualities—without external contributions—would suffice to explain and justify human characteristics. Hence, we founded an anthropocentric anthropology rooted in the ideas of either deficiency or emanation. On the one hand, the hypothesis of human phylogenetic incompleteness (Gehlen 1988) explained the anthropopoietic process as compensation for our scarce natural equipment; on the other, the hypothesis of emanation (Hume 1956) posited that human imaginative and creative talents were a means of fulfilling and creating needs that went beyond the instinctual dimension of animals. Anthropocentric anthropology separates the human from all external co-factorial relationships and strives to define a triumphant and emancipatory path of self-determination, in which the human emerges as the sole craftsman of the cultural processes that were realized in the course of history.

Adopting a posthuman awareness means letting go of the claim that the human was founded *iuxta propria principia* (Terrosi 1997). In other words, we have come to a realization whose consequences are not merely descriptive, but radically modify our philosophical principles. We must gain full awareness of our ontological non-self-sufficiency. The scope of this statement is absolutely revolutionary. The presuppositions of ontological separation collapse along with the idea that, in order to understand and assert itself, the human needs to withdraw into itself. I have referred to otherness as ontological symbionts by drawing on ecology and physiology, disciplines that reveal how every living organism is a conglomeration of entities, a kind of ecosystem or holobiont (Margulis 1999). In biology there is no such thing as metabolic self-sufficiency. Similarly, I believe that there is no solipsistic ontopoiesis, because human ontology has developed and manifested, it has expressed itself, through referential symbionts capable of offering a dimensional contribution.

The posthuman epoch demands profound transformations: we cannot simply limit ourselves to observing the cultural and technical dimension of the human. We must emphasize how ontologically dependent on the nonhuman we are. Hence the need to acknowledge that external entities are not mere resources or tools. The posthuman era is a project that calls for an extremely complex philosophical commitment, because the stakes are high: escaping the apparently comfortable trap of anthropocentrism. Gaining consciousness of our hybrid condition is a process fraught with consequences. It requires surpassing anthropocentrism (Descola 2005) as a reading of human ontology. On the one hand, we need to charter new paths to interpret the human and, on the other, lay the foundations of a new alliance with otherness.

The posthuman rejects: autarky, the assumption that everything is the product of human ingenuity, an ontology based on isolation and a discourse shut up in the cloister of the human dimension (Descartes 1951). What prevents us from gaining awareness of the posthuman condition is not the realization that we have a surplus of cultural apparatuses within and around us but the loneliness and arrogance inherent in our claim for autarky. The error or, if you like, the impediment, lies in this closure. We recognize that a solipsistic abyss goes hand in hand with the *cogito* of Descartes and the transcendental of Kant as soon as we start to realize that we have an inherent—inner—need to open up to the world. Thinking in terms of ontological non-self-sufficiency means admitting that no emergence is possible outside of dialogue.

It is in dialogue that I find confirmation, that is, it is the otherness that speaks of me. If we were lost in space, like astronauts shipwrecked in the obscure immensity of the universe, our existence would lose its meaning, not because we would be unable to doubt nor because of lack of spatial and temporal references, but because we would be excluded from the dialogical feast. Dialogue molds us from the moment we are born, first through our mother and then through infinite referents who continue to dialogue with us, even if they are far away. Posthumanism replaces *cogito, ergo sum* with *dialogo, ergo sum*, namely with a self-awareness that is directly proportional to our dialogical skills.

## THIRD THESIS

For humanism, the human condition is the result of a natural ranklessness that allows for and requires compensation through culture; in the posthumanist view, it is precisely the wealth of phylogenetic a prioris that favors cultural expression.

Humanism explains human predicates in two ways: (i) through ranklessness or the human's inability to be ranked, which is conveyed through a number of metaphors that have emerged during history, such as deficiency, exuberance, larvality, ancestrality and revertibility. According to this hypothesis, the human lacks the attributes of other animals and, therefore, needs external support or "crutches" (Herder 1772); and (ii) the hypothesis of emanation, which considers all cultural products expressions—*ex-primo*: to bring out—of preexisting contents. The anthropopoetic development, in other words, is like unrolling an already written scroll, even though, according to some interpretations, the productivity of human nature is also the fruit of creativity and imagination. Thanks to its imaginative surplus and plasticity, human nature is seen as germinative and creative (Bergson 1907).

Both approaches are attempts to explain the human in terms of self-withdrawal and self-investigation. Their fundamental error is considering the human an ontologically separate entity, self-contained and self-accountable. Their differences notwithstanding, both share the assumption that the human condition can be explained by limiting the study to the human. Obviously, for the human to come center stage, it must lose all connections with the biosphere, which is considered merely a resource. For the human to burst on the animal scene and stand out as the nonanimal par excellence (Heidegger 1992), it needs to preserve an implicit purity, a distinctive trait, which makes it special and nonspecific.

Humanism sees in *natura naturans* a set of determining and binding laws, conflicting with human freedom. Accordingly, *natura naturata* is bound to phenomenological necessity, devoid of will or choice. To claim a space for itself (Kant 2012), the human needs either imagination or insufficient natural equipment. The latter refers to the Epimethean deficit described from Plato to Arnold Gehlen, while imagination is a characteristic that has always been used to justify our separation from other species. What it ultimately comes down to is that both culture and anthropopoiesis can be explained without involving co-factorial entities. Posthumanist philosophy questions this assumption by taking a radically different stand. Human nature remains fundamental to understanding anthropopoiesis. Yet its predicates are neither compensation nor emanation: they are *copula*. The strong human mimetic and epimeletic traitsand its capacity for empathy make it open toward otherness. By virtue of its phylogenetic attributes, therefore, the human is predisposed to hybridization.

From several viewpoints therefore the hypothesis of incompleteness seems to be totally unfounded. First of all, it mistakes an ex post perception of scarcity—due to our habituation to techno-mediatic standards or to the convivial and affiliative dimension of the cultural apparatus—for an original deficiency. This explanatory fallacy is anything but irrelevant, because it takes for causes what, in fact, are the consequences of anthropopoiesis. From a morpho-physiological viewpoint, it also makes no sense to consider the human incomplete, larval, or unspecialized. By comparison with our anthropoid cousins, our anatomical specializations suffice to disprove the hypothesis of ancestrality and incompleteness. Bipedalism has reshaped the entire structure of our bones and muscles, modifying its whole architecture: the shape of the spinal column, the foot, the pelvic structure, and the foranem magnum, to name just a few (Leakey 1994). Moreover, our mimetic apparatuses have also transformed: our oropharyngeal structure has changed, the splanchno- to neurocranium ratio has been remodeled and the larynx repositioned.

There is another fundamental aspect to take into consideration. We cannot rigidly separate the sciences from the humanities in posthumanist philosophy. This division has made it impossible for the two domains to communicate with each other, which explains why there are still philosophers who claim the incompleteness of *Homo sapiens*. Without going into the anatomo-functional and ethological details that prove how hyperspecialized our species is (Eibl Eibesfeldt 1989), we just need to consider the striking differences in the whole sexual dimension: the loss of the penis bone, penis enlargement, estrus no longer apparent and the regression of the Jacobson's organ—even from the viewpoint of evolutionary theory the hypothesis of a negative selection cannot hold. Any process of niche construction, which is what comes closest to the cultural dimension, produces a specialization, namely a selective shift: this means that the selective pressure on certain organs shifts to one on different organs. The selective parameters remain those of sexual competition and competition for resources.

Undoubtedly, the generative model seems to be more consistent with a view of culture as a creative act. Yet it has a fault, which is common among humanists: the idea of an opposition between phylogenetic heritage and cultural emergence (Scheler 2009). Posthumanism goes beyond this antinomy, which is a consequence of the archetypal human-animal dualism. Moderate perspectives, which take evolutionism inside the humanist paradigm (Tort 2004), associate culture with situations in which the phylogenetic heritage is less clear-cut and more possibilistic. For posthumanism there is no inverse proportionality between the innate and the learned, between what is phylogenetically inherited and what can be acquired. There is, in fact, direct proportionality. It is only within a notion of nature as extension that the innate can be considered automatism and viewed in terms of expressive determinism: its explanatory model derives from the dualist assumption (Morin 1974). By overcoming dualism, posthumanism supplants these oppositions and transforms the innate into a copulative factor, an ontopoietic flywheel: greater innate endowments allow and demand from the species a greater intellectual effort and a greater range of creative possibilities or cultural emergences.

It would be wrong to assume that the products of human creativity have come about in isolation and autarky. This is a key point in posthumanist critique. The complexity of human nature does not thwart but amplifies our dialogical skills (Buber 1984). We could argue that dialogue—in posthumanism, the heart of ontopoiesis—is the most explicit and productive quality of the human. Yet the concept of dialogue needs to be clarified, because it is not limited to verbal language. Dialogue is about *mimesis*, care, identification, empathy, self-recognition through the other, exchange, transaction, play, epiphany, possession, and ecstasy. To have a propensity for dialogue means, therefore, being brought to (inclined to), and supported by (endowed with), reciprocation, hybridization, interchange, and alliance. Having a copulative nature implies a willingness to open up (Hartmann 1942), susceptibility to external influences, robustness of introjection, and plasticity in the representation of otherness. It indicates a vocation for hospitality, in the double Latin meaning of hosting otherness and being hosted by it. Human nature is catalytic.

In order to understand the human propensity for dialogue, we need to consider the human brain's potential of neurobiological wiring. But we also cannot forget the role played by two particular copulative coordinates: (i) emotional sentience, enhanced by the development of the parental and social dimension in humans; and (ii) desire as a motivational drive, which makes humans particularly curious of their surroundings. We cannot appreciate our imaginative skills if we do not realize that our sensitive and desiring nature makes us immerse profoundly in phenomenal reality and transform it into epiphanic reality. Anthropological testimonies hardly leave any doubt of the importance played by biodiversity in the development of our first cultural models (Ingold 1992). By reversing Plato's formulation, we could argue that many of the human cultural models are shadows of phenomenal reality.

The human is naturally directed toward relational creativity. This does not deprive it of its imaginative and representational prerogatives. In fact, creativity and imagination are acquired and enhanced through the encounter, and often the partnership, with otherness. A participative and bio-mediated imagination paves the way to a symbiotic ontology: otherness is a cofactor of cultural production. The conclusion we can draw is that it is typical of the human phylogenetic legacy to create a *convivium*, a gathering of contaminating contributors from which culture emerges. Therefore, we can also argue that cultures are the historical and geographical expressions of the relationships established by different populations. Culture neither detaches itself from nature, nor mortifies it. Culture is rather a conjugation of inherited features. Similarly, nature is not a limitation, but a flywheel of culture: because of human copulative potential the entire patrimony of the biosphere can offer us creative inspiration.

Acknowledging human phylogenetic history also means acknowledging that we are a body, a fully integrated organismic entity endowed with distinctive systemic characteristics. Each trait of and in the body has coevolved with the others. Hence, we cannot think of our organs as modifiable modular entities (Macrì 1996). Our organism is a *unicum*, not an assemblage. It is a system with a certain margin of flexibility—a degree of plasticity—but that is not fully moldable. One of the consequences of humanism, is undoubtedly a devaluation of the body: cartesian *res extensa* is a mathematical container; for other authors, the body is mere disposable remains. In the hyperhumanist groove of transhumanism, the body is an obsolete entity, the anachronistic legacy of an animal condition that we need to relinquish in order to achieve emancipation. Propositions that view death as a disease still awaiting a cure, or the mind as a potentially transferrable package of information differ strikingly from posthumanism (More 2013).

Posthumanist philosophy does not seek to be a utopia but a Eutopia, a reflection upon the relationships that sustain the human too. It has its feet firmly on the ground; it is profoundly rooted in the concreteness of nature. If posthumanism considers the spiritual verticalization of the early humanists an error fraught with anthropocentric accidents, it is even more critical toward the transhumanist ellipses as means of mass distraction, because they take our attention away from the actual ecological problem by presuming that the human can save itself autonomously. Keeping one's feet firmly on the ground means understanding that there is a difference between an intervention of somatic enlargement and one of somatic reconfiguration, which is the chimera pursued by transhumanism (Dublin 1989). Precisely because the body is an organism and not an assemblage of parts, it cannot be reprogrammed. We can certainly intervene to amend particular dysfunctions and enhance performance—this is already being done and, most likely, similar operations will become more and more profound in the future; however, it will be impossible to exit the body and find anything that, however vaguely, resembles the human.

## FOURTH THESIS

In the humanist conception, animality is a counter-category against which we can derive the predicative profile of the human; in the posthumanist view, however, animality is a meta-predicative—inclusive—condition, from which the human can be derived by phylogenetic conjugation.

Closely related to the issue of human nature is the broad topic of animality, one of the breaking points with the humanist approach. It is what sets humanist thinkers apart from posthumanist ones (Cimatti 2013). Animality is a pivotal issue that allows us to distinguish the two approaches. Posthumanism reverses the meaning of animality just as it did that of human nature. In the humanist tradition, the animal represents the dark mirror, the quintessence of the stranger, the antinomy of the human, the more or less trivialized entity from which the human distances itself ontologically. Traditionally, to discuss animality means to treat a subject that relates to the human only by opposition (Battaglia 1993). This is an issue of utmost concern to posthumanism: it is the basis of the ontological errors of humanism. This issue is high on the agenda of challenges and problems that the posthuman era will be called upon to solve.

Animality is the ontological dimension that provides the starting point for understanding our existential condition, making peace with ourselves and building a new alliance with the biosphere, a trans-specific feast, as it were. Our reflection, however, differs from the discussions conducted in the last decades of the twentieth century around the critique on ethical anthropocentrism and speciesism (Singer 1975). Even though these discussions issued a warning against the exclusion of other species from our moral framework, they did not question our ontological position and consequently did not gear the debate toward the topic of animality. Ethical anthropocentrism was condemned, whereas ontological anthropocentrism was not. The anti-speciesism of the last century was still ensconced in the humanist paradigm. In fact, similarities between humans and other species, such as sentience, were recognized. These attributes bestowed moral value upon the other species. If we analyze the arguments advanced by Peter Singer, the greatest exponent of anti-speciesism, we recognize a line of continuity with Jeremy Bentham's utilitarian humanism.

For posthumanism, the ethical issue of our relationships with other species must follow a metamorphosis in animal ontology. Anti-speciesism lacks foundation if it remains adherent to the humanist framework: it is a contradiction that is always on the point of collapsing into anthropomorphism. Any form of attention for nonhuman animals will be interpreted as a concession, most often arbitrary, to sit at the table of human rights. This metaphor simply suggests that the audience that can benefit from rights that are considered universal by humanist criteria has enlarged. If we continue to believe that we are nonanimals, however, we create two conflicting and non-dialogical domains. Any reference to animal otherness will be faltering and at risk of being interpreted as a projection, if not a *petitio principii*. What posthumanism questions is not the exclusion of other entities from the universal but the very possibility of a universal based on anthropocentrism.

Animality is the core of the discussion as well as the litmus test that distinguishes humanism from posthumanism. Ontological anthropocentrism is brought to the fore of our discussion, because it is the theoretical framework that considers animality a benchmark for the definition of the human, rather than a dimension that includes the human. In a humanist perspective, the human is present when the animal is absent, whereas in a posthumanist perspective, the human is a conjugation of the animal. For the humanist, animal predicates are opposite to human ones; for the posthumanist, animality is a meta-predicative condition that we cannot exit but only conjugate in particular ways. The animal is not "other-than-me," but it is the "heterospecific," the animal of another species. I cannot say "the animal that is in me," but rather "the animal that I am," because every single expression of mine is an animal expression. I cannot argue that humans differ from animals, because, since animality is meta-predicative, there is no animal counterpart.

As Darwinian evolutionism is not simply a transformist theory but is based on a common progeny, it posed a challenge to philosophy (Mayr 1985): essentialism collapsed by revealing the existence of shared predicates between species. This challenge has largely been ignored. Darwin was twisted to suit a humanist framework, and the questions that remained open about the humanist paradigm were simply swept under the carpet. Speaking only of adaptation rather than of continuity within the differences means ignoring what an ontology based on shared levels is, such as being mammals or being primates. Some less anthropocentric interpreters use the contradictory term "quantitative difference" because they are unable to get rid of humanist parameters. In order to make philosophy and evolutionary biology consistent with each other, we need to understand what the obstacles that stand in the way of ontological continuity are. These obstacles are what induce us to hold on to the essentialism of the classic or creationist tradition. We are witnessing a bizarre interpretative schizophrenia: while science posits a constitutive continuity in the animal condition, philosophy still discusses humans *versus* animals.

The fact that for humans all nonhuman species tend to merge under a single and unified category opposed to their own can be understood (and excused) as the result of a perspective bias—the same one that made the Hellenes say that foreigners were all barbarians (*oi barbaroi*): the distance from otherness is emphasized while the differences between othernesses are leveled. Therefore, despite being close to the human, the chimpanzee falls under the same category as other species, even phylogenetically remote ones, such as the jellyfish. This oppositional categorization of animality is a focal error that a scientist and a philosopher should know how to correct (Christen 2011). If this does not happen it is because the distortion is functional: it provides the very platform of humanist thought, which needs a dark backdrop (the animal) to show how bright the human is. In addition, we need to widen the gap between us and the other species and to level the differences between them in order to legitimize our being special. Perspective bias is an evaluation fallacy, the theory of human specialness is an ideology.

We can identify two anthropocentric matrices. The first one is our point of view, which, as mentioned about biases, is fallible. The anthropocentric perspective can be described as a sort of species egocentrism: it does not offer a neutral view of the world, but one that is biased and inevitably distorted. The second matrix can be called ideological anthropocentrism; it tries to create two different ontological domains, a human and an animal one; despite admitting that there are significant differences between animal species, ideological anthropocentrism positions the animal far away from the human. These two matrices do not overlap, yet they mutually influence each other. On the one hand, anthropocentric ideology does not stimulate any corrections to the distortions of the anthropocentric perspective; on the other hand, these distortions support ideological anthropocentrism.

It is often argued that the very fact of being human condemns us to an anthropocentric perspective. This is a gross error because, just as we can mitigate egocentrism by building relationships, we can similarly amend some of the most common errors of an intuitive vision of reality. We might never be able to fully abandon our anthropocentric perspective but we can certainly temper it through our relationship with nonhuman otherness. Also thanks to technoscience (Bachelard 2002), especially information technology, we can correct some interpretative distortions caused by our intuition. A first step in this direction might consist in tempering the importance of our perspective by becoming aware that a plurality of world perspectives exists, as many as there are species on this planet. As a matter of fact, what really prevents us from leaving anthropocentrism behind has a lot to do with a precise philosophical approach that has sustained the humanistic tradition and become an ideology.

We might also think that the animal otherness, the heterospecific, lies worlds apart from us from an ontological viewpoint, so that we cannot say anything meaningful about it (Nagel 1974). This sort of *epoche* happened several times in history: its remoteness from the human transformed the animal otherness into an alien entity, whose very distance made it unknowable, as suggested by the aphorism *ignoramus et ignorabimus*. Jakob von Uexkull's concept of *Umwelt* (von Uexkull 1956) could be interpreted, against the author's intentions, as a sort of monad: each species occupies a dimension inaccessible to otherness. Once again, this temptation derives from an essentialist approach that radically clashes with Darwinian evolutionism and with the idea of a common progeny. Significantly, posthumanist philosophy is largely indebted to the ideas of the English naturalist. This is the reason why essentialism is like a straitjacket that still hinders a thorough understanding of evolutionism.

The evolutionist perspective assumes that most of the characteristics of a species evolved prior to the origin of the species itself. Accordingly, animals from different species have a good number of shared characteristics (homologies), directly proportioned to their phylogenetic proximity (Geoffroy Saint-Hilaire 1829). For obvious reasons, humans and chimpanzees share an enormous number of predicates, which evolved before the two species separated. Different *Umwelten* are not disjointed monads, but rather entities that reveal extensive overlaps, especially when animals are related to one other, such as humans and other primates or, more generally, mammals. There are also large areas of overlap if we widen our perspective to include birds and reptiles; they then progressively decrease, but never disappear (Shubin 2009). The heterospecific, therefore, is not an alien, but an entity I can understand, at least partially, and with whom I can identify without the risk of anthropomorphizing. This is not because I am attributing human traits to the heterospecific but simply because I recognize that we have characteristics in common.

I would like to underline that it is not just a matter of understanding or feeling close to the heterospecific but of viewing human ontology from a different perspective altogether, of considering it the expression of multiple shared levels rather than an essence. Human ontology is not a circle of which we can measure the boundaries, but the result of overlapping levels and of predicates shared with otherness. Its image radically changes. Human ontology cannot be grasped at first glance (Merleau-Ponty 2013), something which would require geometric solidity. Like a slide under the microscope, human ontology must be examined in its various constituent layers. These layers, however, do not belong to it, since none of the predicates expressed by the human can be considered its own. Once more, the human is neither the measure of the world nor of itself. Interrogating animality in its meta-predicative expressions, therefore, is like a preparatory seminar for an investigation into the human after we have disposed of the oppositional approach.

Animality is a condition that can find different conjugations both in species and individuals. More correctly, animality conjugates the singularity of experience. However, it is also a recapitulation of past times, it is resonance, it is the expression of the *Having-been-there-before* in the world of the *Being-embodied* of the individual. This horizontal continuity in time and in different conjugations profoundly transforms our view of existence. Meta-predicative continuity means that we share an a priori with other species. We have not been thrown into the world but, on the contrary, we are persisting in the world in a flux of transformations. There is a continuity of being in the flowing of bodies that connects us to the rest of the animal world. Our bodies and the bodies of other animals are connected along the line of evolution. Posthumanism strives to mend the rift between philosophy and Darwinian thought by enhancing the concept of animality as a shared ontological condition. Our awareness of being animals paves the way for a profoundly different vision of our relationship with otherness: a community vision, based on mutual gaze, alliance and existence. The human no longer finds itself in being separate from animality but in adhering to it, in the warmth of animal sentience and animal desire. Posthumanist poetics draws inspiration from this new alliance. This is apparent, for example, in theriomorphic art (Vergine and Verzetti 2004).

Posthumanism rejects any form of anthropocentrism and considers the animal condition a meta-predicative dimension that comprehends all species, including the human one. Our dimension is therefore based on the meta-predicates we share with other species but that we have conjugated in distinctive ways. These meta-predicates include subjectivity, creativity, desire and the peripatetic condition, namely mobility in the world. Desire projects the individual into the world and makes it a mobile or peripatetic entity, an entity that searches the world in order to find opportunities to express itself. Walking, flying, and swimming are consequences, not expressions, of this mobility. While meta-predicates are qualities shared by all animals, predicates are the different conjugations of the animal condition: they enable us to tell one species from another. The dialectic human versus animal is impossible to maintain, because the animal includes the human; we would otherwise be comparing a predicative profile (the human) with a meta-predicative one (the animal), which includes the former. The human can only be evaluated on the basis of characteristics that are applicable to it; it cannot be compared to the dimension that includes it. In other words, it can be compared to another species but not to animality. There is not a human versus animal ratio; to formulate a comparison in whatever form—including the claim that "humans differ from animals in . . ."—is a serious mistake.

What we observe in humanism is that animality remains a contralateral condition, to the point that the human and the animal start a dance of oppositions, which creates an actual anthropological machine (Agamben 2004). If the human has free will, the animal is limited by its rank; if the former has reason, the latter has instinct; if one is self-determining, the other one is determined, and so on and so forth along a logic of contrasting dichotomies that elevate the human while demeaning the animal. In this perspective, the human is not a particular animal but a nonanimal, as posited by Martin Heidegger in defining the ontological difference. We can obviously hold on to this view only as long as the essentialist paradigm survives, and as long as we keep on considering the technological and cultural apparatus something which is external to us, like a garment. If we adopt an ontology based on a thousand plateaus of sharing and if we acknowledge that animality is a meta-predicative dimension that can be conjugated we take a revolutionary step in our interpretation of anthropopoiesis.

## *Chapter 2*

# The Relational Dimension of the Human

Posthumanism is a relational philosophy that considers the human condition the expression of the different relationships we have built since the Paleolithic times (Descola 2014). The theses presented in this chapter argue that it is impossible to understand the human dimension by studying the human from within, because the human dimension is the result of infinite hybridizations. The chapter challenges the essentialist view of the human, the idea that human predicates are autarkic expressions originating from *Homo sapiens*. It rejects the solipsistic view of the human and supports a relational interpretation, based on hybridity with otherness. This is the basis of posthumanist philosophy: the process of *anthropopoiesis* (Remotti 2012), namely the development of a cultural dimension, is not separative but connective, not vertical but horizontal. The big difference with the humanist paradigm is a yearning for conviviality with the biosphere, a rejection of the modern tendency to consider the human self-contained.

## FIFTH THESIS

The humanist perspective is predominantly essentialist and views entities as interacting but impervious to each other; the posthumanist perspective is eco-ontological, as it considers predicates expressions of hybridization.

Posthumanist philosophy takes up the challenges posed by the twenty-first-century scientific evolution as well as by the epistemological changes that occurred in the past century in order to develop a new ontology based on hybridity, that is to say an *eco-ontology* (Marchesini 2018). Posthumanism disputes the essentialist perspective and views phenomena as hybrids—originating as hybrids and expressing themselves as hybrids. Posthumanism rejects mainly the qualities of purity (Douglas 1966) and authenticity but also the processes that favor and support them, such as purification, purging, autarky, autonomy, autopoiesis and adherence to an essence. For eco-ontology, entities cannot be separate or self-sufficient in their origins, characteristics or expressions (Scott and Epel 2015). The reasons every single phenomenon is a hybrid are the following: (i) it originates from other entities, inheriting shared traits; (ii) it is part of a historical flow of gemmation processes so it shares numerous levels with other entities; (iii) it absorbs features from the outside world and transforms them into a prioris: the entity therefore tells us about its own particular and specific world; (iv) it is predisposed to openness (heteronomy) through symbioses. Symbiosis is what makes the expression an emergence, a new trait resulting from the combination and synthesis of (at least two) elements.

According to essentialism, entities conform to their essence, they own discrete and exclusive predicates. Even when interacting with the world, entities remain essentially impervious to it. For essentialism entities have been thrown into a foreign world (Heidegger 1927) that is either suspicious or alienating. Viewing the predicate as a hybrid reverses all these assumptions. For posthumanism, the hybrid is how an entity emerges from flux and manifests itself. Because each entity shares qualities with other entities, it is impossible to extract its qualities from its preceding (constitutive) and phenomenological (expressive) proximities. In other words, I cannot understand the qualities of an entity if I do not understand the relationships that formed it as well as those that permit its expression. Nothing can ever be considered pure, authentic, perfect, proper or separated from anything else (Margulis 1999). This should not lead us to zero out the qualities of an entity, however, but to change how we interpret their emergence. Sharing is the logic behind the emergence of predicates, because a predicate both shows a correlation with the world and needs external contributions to manifest.

These three points (sharing, correlation with the world, and contribution of the world) mark a profound division between humanism and posthumanism. The logic of sharing not only questions the *principium individuationis* (Leibniz 1980), but also makes the entity an expression of conviviality. The entity shares different levels with other entities and multiple predicative belongings. For example, being human does not only mean belonging to the species *Homo sapiens*, but also belonging to the order of primates, the class of mammals, the row of vertebrates, and so on. Accordingly, no entity can be considered a monad, not simply because it communicates with the external world, but also because it partakes of something beyond; it is always predicatively a "we." Overlapping *Umwelten* provide an example. Let us think of what is usually considered a human attribute, such as maternal love. What happens when we realize that this distinctive human trait actually predates the human, because it is shared with nonhuman entities? How shall we consider it?

If an entity emerges by serial gemmation but its emergence is also related to the world, the entity not only anticipates the world but is also configured so as to receive it (Campbell 1974). It requires the world's input, like the gills of a fish require a liquid environment and the wings of a bird the resistance of air. Predicates come about by copula, not by emanation; they are never pure because the entity is made to hybridize. Therefore, the world is not a stage but a cocreator of predicates. In posthumanist philosophy, phenomena are the fruits of relationships, and nothing exists outside relationships. However, the notion of the relationship itself changes: it is no longer the interaction between impermeable entities that influence each other without affecting their identity matrix. An entity outside hybridization is ontologically impossible (Marchesini 2018). To speak of mere influence would also be an epistemological error. The hybrid is an entity that assimilates and subsumes the entire flux of events that preceded and are related to it. Because of its constitutive plurality, therefore, the hybrid is always a singularity.

Singularity, however, means neither complete difference nor chance. It means continuity and transition, conviviality and emergence. The hybrid is not a human who has been technologically empowered. And it is not the retrieval of concealed animality either. So, the hybrid is neither represented by the cyborg nor by the theriomorph. The hybrid reminds us of an existential condition that breaks the wall of isolation along with the despair and pessimism that follows the realization of having been thrown into the world. The hybrid does not share the destiny of the castaway in the humanist tradition. Posthumanism, in fact, is a philosophy of liberation (Braidotti 2019), albeit not one envisioning the breakup of bonds and the emancipation of the individual. It is a philosophy based on a logic of strong participation. It is a reminder that every single action we do in the world is something that also affects us ourselves—regardless of deontologies, karma or other forms of transcendence. The hybrid must first of all come to terms with its *being-for-the-world*, consistently with its own convivial and emerging singularity on the basis of its own possibilities.

The technopoietic pervasiveness of our time—digital immersion, the IT revolution, electronics, and biosynthesis—is not what makes hybridization the major ontopoietic factor in the posthuman era. In fact, these transformations may have rendered the connotations of hybridity more evident (Sorgner 2021). Because of the singularity the human has achieved today, it may have gained a greater awareness of its connections and have fortified its *being-for-the-world*. Applied to an epoch of transition and becoming like today's, the term "posthuman" does not mean that we are transforming ourselves into something other than human, or that we are abandoning the human condition altogether. It means we are acquiring a more mature and responsible awareness of our ontological non-self-sufficiency. We need to realize that we must overcome boundaries and make ontology a fluid condition through which multiple connecting vessels let existential singularity emerge in a logic of continuous mutation.

The hybrid therefore neither lives at the margins nor represents the different; it denies the existence of real boundaries and of absolute otherness. To become posthuman therefore means to accept the Dionysian as a choral flux that permeates existence, while the Apollonian is an artefact (Nietzsche 1995), a temporary but real and concrete shape that has emerged from its virtual condition. Form is therefore an expression of singularity; it is not absolute, but subsumptive, correlated, contingent, and transitory. In its convivial existence, the horizontality of the posthuman opposes the verticality of the human: The latter yearns to ascend as much as the former aspires to be part of something. The posthuman is hybrid not because it is modified by technological prostheses or immersed in digital realities, or because it is a neuromantic (Gibson 1984) inhabitant of another world, or because it is theriomorphized by operations of biosynthesis. The posthuman is hybrid because of its different relationship with the world: one no longer based on oppression or on the illusion of inhabiting a presumed Eden or a prison of the soul. The posthuman is hybrid because it is longing to rediscover the conviviality of life.

The author who mostly influenced the advent of posthumanism is probably Charles Darwin. Even though he did not formulate a philosophical treatise on the subject of relational ontology, his theory of a common kinship—that species share common predicates—and of populational nominalism eroded the foundations of essentialism (Mayr 1982). The fact that an entity is relatively different from others is no longer considered a degeneration from a perfect essence but a process of productive mutation. Evolutionary biology buried essentialism and offered a new version of the origin of human predication. To date, however, the latter remains solely descriptive. Therefore, we need to find a new philosophical system that may clarify all the phases of the predicative event and apply it to ontology. A great change of perspective comes from the acknowledgment that no phenomenon can be exclusive and autarkic (Patocka 2016). Ontology, however, still suffers from the influence of Platonism and Aristotelism: predicates are identified by studying the human only. In my opinion, this is the core problem of humanist philosophy.

This does not mean that we should deny the existence of inherent characters. Rather, it means: (i) in terms of constitution, that these characters are the fruit of the contribution of other entities and of correlations with the biosphere; and (ii) in terms of expression, that they emerge as a consequence of the relationship envisaged by the inherence. If we consider the convergence between the new models of extended evolutionary synthesis (Pigliucci and Müller 2010) in biology and of complexity theory in ecology (Bocchi and Ceruti 2007), we notice a profound change in our way of interpreting life phenomenology (Tymieniecka 2007). An organism used to be thought of as the interactive manifestation of an entity and the product of a recipe that defined its characteristics with great precision. Even though it is undoubtable that all entities have inherences, we need to understand how they come about and express themselves. It is here that posthumanist philosophy makes a difference. To reason in eco-ontological terms means: (i) to consider each predicate the co-factorial result of the entity's inherences and of external contributions; (ii) to consider inherences the outcome of previous dialogical processes and of their introjected relationships; (iii) to view all inherences as a virtual space to be configured, a flywheel of hybridizations; (iv) to conceive virtuality as a field with both limits and opportunities that define a range of possible conjugations, rather than a *tabula rasa*; (v) to interpret virtuality as being predisposed to hybridization and as having expectations of the real world.

## SIXTH THESIS

In order to understand the posthuman vision, it is necessary to shift from a conception of self-sufficiency and autopoiesis to a relational conception of the phenomenology of life, leading to a shared, heteronomic, and sympoietic reading.

If we analyze the biological organism as a separate entity, we cannot understand how life phenomenology unfolds at multiple relational levels. Each organism presents itself in multiple ways: any of the images we derive from it, albeit not arbitrary, is always partial and transitory. We cannot reach an entity's core because each level of the entity's expression is the result of previous relationships and of the particular relationship that is expressing it. This is the reason why an autarkic and autopoietic (Haraway 2016) view of life phenomenology is unacceptable. Posthumanism replaces the idea of a self-confined and self-contained entity with the idea that an entity is characterized by thresholds of relationships and different levels of sharing. This change in perspective reveals that, in the here and now, the subject, affected by the singularity of its status, is in a constellation of relationships with its proximity; if we continue to search, we will discover a cultural dimension, resulting from the relationship with other cultures, and then, a human condition connected with other species, and so on and so forth along the relational plane.

The humanist error derives from our need to circumscribe the entity in order to pin it down to a precise and clear-cut role. In the case of the human, this role is that of ruler of the biosphere. What posthumanism seeks to expose is the fact that whenever I circumscribe an entity in order to isolate it, I am forcing it into a discrete role. In other words, the error lies in taking the characteristics of one particular level for the essence of the entity. What is arbitrary is not so much the image we get of the entity as the representative level we choose for it and the hypostatic level we attribute to it. What the relational phenomenology of life reveals can be summarized in three words: (i) heteronomy or predisposition to relationships; this is the core principle of all inherences: the entity anticipates relationships or is configured so as to be formatted by them; (ii) sympoiesis or the predicative manifestation of the entity in various ways, depending on the relationships it establishes; this makes it impossible to separate the entity, a phenomenological expression, from the ontopoietic symbionts that define it—for instance, a genotype produces different phenotypes depending on the environment; and (iii) sharing, belonging to multiple levels: rather than exclusive to one entity, an inherence can appear in all the other entities that are positioned on the same level. For example, when I recognize that maternal love is shared by all mammals, I stop considering it a human characteristic only and begin to view it as a characteristic of mammals.

Let us begin with the last point. This will help us understand animality as a metapredicate and identity as participation and sharing. As observed, inherences are contents; they are what an entity brings into its dialogue with the world. They are not essences but the introjected results of previous dialogues. They do not actually distinguish a particular entity or separate it from others; they just place it on a relational level with multiple others: for example, humans, mammals, or tetrapods. We must therefore view the entity as a fractal of belongings rather than something separate and ontologically opposed to otherness, simply because we share levels even with the living beings that are the most remote from us. The predicative specification of the entity, therefore, cannot be a dialectical operation that sets the entity in opposition to others. In fact, it must result from the analysis of the different levels shared: each specification is never absolute but depends on the chosen level. Most of all, it is impossible to consider the inherence of an entity an essence, something pure; it simply reveals that the entity belongs to multiple levels.

Essentialism is the legacy of the creationist and transcendental perspective, which frames the entity within a network of harmonious metaphysical relationships. In the essentialist perspective, there are only two types of relationships, neither of which really envisages hybridization: an entity may be defined by either opposition or by the type of relationship it establishes with other entities. The former view is disjunctive, based on self-sufficiency and autopoiesis. It assumes the existence of a full-fledged entity that owns its predicates and addresses the world like a passive host. The entity is autonomous, self-sufficient, different from the real that it dominates. This relationship may be described in terms of opposition, of *polemos* (Hegel 2018). The relationship is limited to a comparison. However, there are other cases in which the relationship puts the entity in a predicative position. What defines my sister, for example, is a relationship of parental sharing with me. Our parents define the relationship I have with her: she *is* my sister. She is defined by being my sister and not a friend or anything else. The attributes of the entity and the type of relationship coincide. Our relationship is something that precedes the making of her subjectivity; it does not affect it at all. My sister might have qualities that are independent from our relationship; it is not our relationship that changes her. This view does not move an inch away from an autopoietic idea of the entity, which remains semipermeable to external influences. Neither of these types of relationships applies to the relational phenomenology of life as I conceive it. When I refer to a constitutive relationship, I mean a relationship that affects and changes the entity.

Let us examine heteronomy (Marchesini 2021). To assume that the entity is predisposed to relationships means reversing the explanatory paradigm of essentialism. This does not mean that the entity is incomplete or deficient, or that it is shapeless. Heteronomy is not a vacuum to be filled, or an amorphous condition that needs to find a shape. Heteronomy conveys the complexity of the entity. The entity needs an external contribution, as if the organism were only just one part of the individual's existence. An individual's existence implies the existence of a co-factorial world. Heteronomy is a kind of ecological heritage (Odling-Smee 1988) that is just as constitutive of the entity as its genetic legacy or its epigenetic marking. We must think of the entity as something that needs to be formatted by the outside. The phenomenological complexity of life relies on the external world and avoids autonomy for the simple reason that without an external contribution, it could not fulfill its own requirements.

Life's heteronomic condition radically twists the principle of essentialism and paves the way for a new life phenomenology. If we hold on to the humanist paradigm, however, we are at high risk of misunderstanding. That it needs to be formatted by the outside does not mean that the entity is incomplete and exposed to the world. Heteronomy is not about completion or fulfillment, because it is not triggered by scarcity, but by redundance. Within the complexity of living phenomena, the entity is never in a shell: it always bubbles out and structures itself internally in order not to be self-sufficient. Heteronomy implies that there are multiple possibilities of conjugation, multiple locks to accommodate specific keys of the world (Tinbergen 1965). By "heteronomy," we assume an entity's predisposition to hybridization, since hybridization is both necessary and inevitable. An entity's ontopoietic traits therefore display a copulative matrix as well as a virtual structure that derive from redundance rather than scarcity. Ultimately, heteronomy is a piece of equipment meant to facilitate predicative hybridization.

Conjugation neither makes up for a deficiency nor gives shape to a formless structure. It means that the entity communicates with the world in order to mold its own characteristics and adapt them to the situations it will need to face. Let us take an example. A cat is equipped with a predatory motivation, an urge to chase and catch small moving entities. This produces a drive toward the world that has a precise orientation (copula): the cat's interest in movement. Moreover, the cat is innately equipped with: (i) gestalt references to identify preys (Köhler 1929); and (ii) a generic choreography to perform and complete a predatory sequence. Both the motivational copula and the innate equipment that enable the cat to recognize and hunt prey are like a model that may be reconfigured. The generic gestaltic indications and the nonspecific choreography, therefore, will turn into a particular search image and a particular hunting pattern, depending on the prey.

Heteronomy is ultimately one of the entity's constituting contents, an inherent character outwardly directed in order to acquire a discrete configuration. Inherence is virtual because it may be configured in many different ways. Inherence does not directly produce predicates but enables their emergence through a dialogue with the world. Inherence is heteronomic because it expects the outside world to format it (Waddington 1959). This has nothing to do with incompleteness, just like the infinitive form of a verb is not incomplete; it contains virtually all the conjugations that the entity's dialogue with the world makes possible. Inherence defines a range of possible conjugations, which are neither virtual nor formless; the form is not imprinted from the outside in a deterministic *tabula rasa* kind of way; it is envisaged by the configurations possible in the virtual model. So, heteronomy pre-constitutes the entity. The entity requires conjugation, because its inherence envisages a hybridizing dialogue, not an emanation.

We finally get to the notion of sympoiesis (Haraway 2016), which in life phenomenology is the opposite of autopoiesis. Science has shown that the individual is similar to an ecosystem: the surface of our body contains a large number of symbionts. Many also live in our gastrointestinal tract. In fact, 90 percent of the cells in our bodies do not contain the human genome. Moreover, many endosymbionts are present in human cells (Scott and Epel 2015). How can we translate these findings into a philosophical framework that may interpret their consequences and the resulting problems? This is a core point of the posthumanist proposal, based on an eco-ontological perspective, namely the acknowledgment that ontopoietic symbionts influence or, rather, coproduce the existential dimension of individuals.

The discrete lives of individuals derive from focusing on a particular level that does not take account of either the preceding level—ontopoietic symbiosis, which brought it about—or the subsequent one—niche relationships. Sympoiesis therefore inaugurates a new awareness: ontology is the building of alliances, which develop different functions at the different levels that make up the entity (Kropotkin 2020). To acknowledge multilevel symbiosis means not only to consider how important it is to protect relationships, but also to be aware of how fragile life is: it needs to be sustained by a network of alliances, each of whose relational intersection is constitutive. Sympoiesis shows how life phenomenology unfolds through flux and participation, it reveals that each level is an interweaving of anastomoses that are impossible to remove. As soon as the human realizes that it is part of an ecosystem, a symbiontic dimension among living beings and a biological assembly rippling through further dimensions of cooperation and inclusion, its existential perspective changes: the humanist view that considers the individual an inhabitant of a space-environment becomes obsolete.

## SEVENTH THESIS

The humanist vision is based on an uncritical celebration of the human, which is regarded as an entity that, precisely because of its neutrality, can aspire to universality; the posthumanist vision adopts conditional trust toward the human, rejecting the idea of a neutral condition.

Freeing the human from all possible shackles in order to celebrate its dignity has been the manifesto of humanism. This has happened through the exaltation of self-determination, ontological plasticity, Prometheanism, anthropocentrism, the power of reason, creativity of spirit, and *egotheism*, the deification of the individual self after the death of God. Without a natural rank or the limits established by Epimethean virtues, the human acquires demiurgic qualities and becomes a totipotent entity (Pico della Mirandola 1942). There are undeniable similarities between transhumanism and early humanism, a humanism yet untouched by the pessimism of the nineteenth and twentieth centuries. We could argue that transhumanism makes a leap from the Enlightenment to the twenty-first century: it celebrates a human empowered by technoscience, an empowerment that the fathers of humanism attributed to the spiritual dimension (Bostrom 2016). It is precisely from these authors, however, that transhumanists draw inspiration. The human is a neutral container to fill with self-defined predicates. Once intrinsic human motivations are zeroed out, however, it is not clear what will direct our metamorphic journey. The potential of technoscience is insufficient to explain the transhuman project.

This view, which is often confused with posthuman philosophy, may at first resemble some sort of uncritical technohypnosis. Its sci-fi kaleidoscope rests upon projects and assumptions that are the same as anthropocentric Prometheanism. Other than the technological ellipses—such as mind-downloading, artificial intelligence, prosthetic enhancement and biosynthesis (Shimasaki 2014)—the novelties of transhumanism are barely noteworthy. They revolve around the trite exploratory impulse toward new existential dimensions, new forms of the old myth of the frontier, the hackneyed issue of immortality, the desire to increase our dominion over the real and to acquire new forms of hedonism, and the usual androcentric complex that interprets machines as compensation for the fact that the males do not give birth. Far too little for a philosophy that aspires to break away from the humanist tradition! Posthumanist philosophy, on the other hand, does not question the human condition but how we interpret it in order to get a more coherent and reliable picture of technopoiesis and technomediation. This new interpretation replaces the model of humanism. What changes is how we consider human nature, our relationship with otherness, the cultural dimension, and technomediation. It is in all these areas that the posthumanist critique thrives and offers new ontological points of reference.

As far as human nature is concerned, posthumanism rejects the ideas of ranklessness and neutrality. The human derives from a specific phylogenetic trajectory, a specialization process. This process has not only radically redefined human anatomy but also conferred to the human mind a precise cognitive and dispositional setting, which influences its way of being in the world (Eibl-Eibesfeldt 1989). We have abandoned the idea of having been thrown into the world and have embraced the idea of *Having-been-there-before*. Not in terms of metempsychosis, though, but in the sense that the memories of our ancestors survive in the a prioris of the individual (Lorenz 1978). The individual enters life not as an alien cast into a foreign world, but with the joy of discovering a reality that resonates in him/her, because he/she is already familiar with it. The image that best depicts our relationship with the world is not that of a newly born baby crying. This is merely the temporary expression of the toil of birth and urge to breathe. The best image is that of a baby smiling when it first meets its mother's eyes. The joy of the infant expresses the pleasure of finding ourselves in the world and realizing that it offers what we had been anticipating, it is the happiness of discovering a coherence between our inner contents and what the world offers.

*A prioris* are, therefore, phylogenetic *a posterioris*; they are a hodgepodge of memories selected by phylogenesis. These settled memories have become an integral part of the individual: he/she owns them in spite of not having experienced them directly. They are a resonant pandemonium of selected memories that subsume the phylogenetic transition terminating in every single individual. These memories were scattered in different points in time and mark important shifts of the phylogenetic chain: we still harbor memories of fish, reptiles and primitive mammals and, further on along the branch of primates, anthropoids and australopithecines. What we are referring to is a distinctive, cognitive way of processing information, of producing hypotheses, of categorizing events, of associating variables, of formulating conclusions, of reaching judgments, of extrapolating metacognitions and memorizing events. Because we do all this in our own, discrete way, we cannot be objective or aspire to become a universal term of reference. Like every other species, we have our own perspective on the world. Our perspective is affected by the selective pressures that have operated on our cognitive apparatus. We are even equipped with a set of predictive laws—so-called naive physics—and we are surprised whenever phenomena do not seem to follow the rules we expect.

However, our actions are mostly influenced by our inner motives, namely our feelings and desires (Marchesini 2021). With few exceptions, humanism has attempted to separate judgment and moral action from feelings by allocating them exclusively to the domain of reason. Results have been disappointing. Our feelings support our moral commitment and help us avoid mechanisms, such as inattention, negligence, carelessness, or even removal, that might compromise our so-called broad morality. We are moral agents not in spite of our feelings, but because we are comforted by them. Reason and feelings mutually support each other and keep the fire of moral commitment burning. We cannot disregard the emotional phenomenology of the human. In fact, we ought to be fully aware of it in order to be able to understand its points of fragility and strength. We ought to forsake exclusively rational ethics, which view moral action as grounded on the non-animality of the human. On the contrary, it is precisely human animality that sustains our way of being moral entities.

Similar observations apply to our *desiring condition*. Desires are infinite because there are infinite desirable things or goals. However, the roots of desire—which define the desiring condition—are to be found in a few motivational matrices that can be conjugated in various ways. For example, humans have a predisposition to collect, namely a tendency to fetch and hoard. It is of course possible to fetch and collect many different things in as many different ways. Hence, motivations are like verbs (to collect, to chase, to explore, to care, to demand, to compete, etc.) that can be conjugated with circumstantial complements. Desire refers to the affective condition (Sartre 2001), to its predicative function. This means that we need to abandon the humanist idea that desire expresses a lack, a scarcity, or a longing for something missing. Desire is the copulative exuberance expressed by the action, in other words the verb. The desiring proactivity of the subject becomes manifest in the conjugation—such as collecting stamps—but its origin is the verbal predicate that characterizes the entity: the human *collects*. The object-result only provides an opportunity to manifest the action.

Desire is languor, a yearning to express oneself. It is something concrete and real, a state of restlessness and trepidation, hence of wealth rather than scarcity or deficiency. Although it has no content, the *desiring condition* is not without specificity (Lacan 2021), its specificity being the verb that defines the action (e.g., to collect). Then, in relating to the world, desire acquires content, because the world offers real targets, objects and things among which desire orientates. In other words, the world is like a field of opportunities where the *desiring condition* can pick and choose on the basis of an individual's nature. If my languor is to collect, I shall be looking for things to collect, such as shells or stamps or antiques or books. But what prompts me to look for shells or stamps or antiques or books is my wish to collect. While desire expresses itself in the action described by the verb, the object-result of desire offers merely an opportunity for expression. To recognize the intrinsic motivation behind human proactivity, therefore, means to unravel the predicative motive behind the conjugated action: you do not desire the object in itself but you desire to express your vocation to collect.

The desiring condition bears implications for human actions: only a handful of motivational verbs apply to the human condition, but they can be conjugated in infinite forms. We have the tendency to think of desire as nonspecific and totally undefined languor. The object of desire, on the other hand, is always subjective and extrinsically concrete. This easily leads us to take the object for the focus of desire: the object becomes the protagonist. However, this mistake redirects us to humanism. Viewing desire as the lack of something that we can find in the external world assumes that the human is imprisoned in the utopia of the indefinite. This is in line with the humanist paradigm of neutrality and ranklessness and with the idea that the human has nothing to do with the animal dimension. On the contrary, if I understand that languor is the need to express my own nature, I cannot think that desire originates outside me, in the world (deficiency) rather than in myself (languor). Desire is, in fact, what drives human actions. It is a principle that, just like the chemical properties of a certain compound, deserves our full attention. It harbors the excellence of human behavior as well as its evils, the springboard for active life as well as the problems of psychopathology. Ignoring these tendencies jeopardizes our capacity to interpret human behavior and our chance of finding adequate remedies to critical situations.

To acknowledge that human nature has cognitive, emotional, and desiring specificities means to reject a universalist view of the human. It means admitting that the human perspective is partial and it also means applying to the mind the same interpretative methodology used for the evaluation of sensory organs or the conformation of the body. Moreover, the point is not to justify all types of behavior as being in line with nature: nature defines expressive predicates and not clear-cut outcomes. For example, humans have a predisposition to collect or to compete, but not to hoard corpses or start wars. To get to know and understand our own nature means to learn what the predicates behind our actions are to become aware of our copulative and elaborative flywheels (e.g., to collect and to compete). This will help us prevent problematic conjugations. If we ignore the tendencies of the human and dwell solely on its conjugations, we lack the resources needed to prevent new problematic conjugations. This is the reason why horrors repeat themselves in history. The same tendencies are always recognizable under the surface.

Humanism has a triumphalist, arguably an acritical, view of the human. This error continues to cost us dearly! It is not that we should give up our trust in the human, embrace absolute pessimism or advocate a fierce anti-humanism to the point of promoting self-extinction (Shepard 1998). Posthumanism discloses a third way out, which avoids both humanist triumphalism and anti-humanist pessimism. For posthumanism, we must keep on trusting the human. In the current circumstances, it is indeed up to us to find a solution and an alliance with the biosphere. Our trust in the human, however, *should remain critical*, always alert to its vulnerabilities, temptations and conjugation errors. Because of its exuberance, human nature is explosive, it is imbued with will but also drenched in a sense of power. It is capable of reaching peaks of beauty but also troughs of wickedness.

## EIGHTH THESIS

The humanist era embraced a project of anthropoplasty, namely the conversion of the biosphere into a by-product of the human, making human labor the sole producer of value; the posthuman era will have to reverse this process.

Anthropocentrism is the great evil of the humanist epoch. It is the rancid fruit of the liberating vindications of the early humanists, which, over the last two centuries, have been radicalized by the whirlwind development of techno-science. The humanist epoch exhibits an ideological anthropocentrism that is different from the anthropocentric perspective. Ideological anthropocentrism does not look at the human just from a human perspective, from a perspective mediated by human perceptive and cognitive organs. In other words, ideological anthropocentrism is not simply the perspective made visible by the human lens. Anthropocentrism becomes an ideology claiming that the human has broken free of the biosphere and acquired dominion over it. In addition, ideological anthropocentrism is encouraged and supported by a burgeoning operational system that fosters our spirit of conquest through technological intoxication. The technopoietic development of the modern age, the European expansion into the American continent and, later on, the beginning of the great voyages of discovery created a particular *zeitgeist*, envisioning mobile frontiers and a magnificent progressive destiny for humanity. Unreserved confidence in the human capacity to increase its prerogatives emerged and consolidated. When the Copernican revolution began to undermine geocentrism, the human need to find a center of gravity in itself and to accelerate the transformation of the world into human metabolites became even stronger.

There is one aspect of anthropocentrism that is not always explored. We prefer to deal with the ontological principle of Vitruvianism or, at the most, the ethical issue of our relationship with other animals (Singer 1975). So we confine ourselves to the theoretical or practical consequences of assuming that we are separate from the biosphere. I would like to draw attention to two elements linked to seventeenth-century humanism: the emerging spirit of conquest and conversion. Both derive from our having lost our inhibitions toward nature and the resulting sense of respect for its sacredness. Inhibition and respect originated from wonder, awe, astonishment and the feeling of the sublime. The development of technologies that amplify human operational potential encourages an arrogance that fosters anthropocentrism. Its first goal is to translate the world into human language (*anthropoplasty*). The humanist epoch desacralizes the biosphere. Massive exploitation of nature is justified by assuming that converting the nonhuman into a human product increases its value (Haraway 2016). We do not only refer to an unlimited use of the biosphere to meet basic human needs, but also to an out-and-out project of conversion that views the transformation of the biosphere into human products as an advancement.

This conversion campaign starts as we begin to consider the sacredness of the biosphere a useless superstition. The biosphere is then reified and becomes mere corporeal substance, Cartesian res extensa available for human usage. As in a campaign of evangelization, the human rushes to transform the telluric and the demonic into the anthropocentric and the spiritual. The nonhuman must be converted. It must acquire the soteriological features of the *imago dei* demanded by the catechesis of anthropoplasty (Harari 2015). Dissenting voices are charged with apostasy, as if they questioned a creed rather than a principle of truth. I do not believe that we can fully understand the humanist project if we refer only to the indiscriminate exploitation of natural resources. The focus of humanism is actually anthropoplasty, the conversion of the biosphere, prejudicially viewed merely as extensive matter, into a product reworked by the human so that the human may appear, in part at least, as something different from nature. While nature has no value, the human product, the fruit of human work, has gained value.

In the humanist era, value and human work overlap. The overall value of the world increases with the amount of work devoted to its conversion. This is why, though formally correct, arguing that anthropocentrism places the human center stage fails to grasp all the implications of the humanist epoch. The conversion machine urges the human not just to exploit, but to convert the biosphere into a by-product of human labor. Machines represent and symbolize human creative power (Rosati 2017), the fruit and the tool of anthropoplasty. Like a virus, the human captures every single cell of the biosphere and imposes its own project of synthesis onto it. Not a single nonhuman will be spared from domestication, nor a single forest from agricultural transformation. The frenzy of conversion is a consequence of how the human views the biosphere: no longer as an Eden to dwell in, but as a mass of raw materials to process.

A forest must become land to cultivate, a hilly area be burnt for pasture, a river channeled, a swamp reclaimed, and a hedge with trees and shrubs felled. A bramble bush is something dark and threatening that must be razed and the sea a large store of fish to be caught. The animal is also classified according to human interests. A denaturing process gradually but progressively strips the animal of otherness and transforms it into an object. If the animal is merely extensive matter, it can likewise be reprogrammed. Every time it happens, this transformation adds value, and I daresay it redeems: hence the term—"reclaim." This process could be trivially baptized "the metamorphosis of the natural into the artificial via human labor." It might even be described in religious terms: the wild is a demoniac presence that must be destroyed or an infidel that must be converted. Such a frenzy finds no explanation outside ideology.

Anthropoplasty is ultimately a dissipation rather than an increase in value from all viewpoints, including, in a long-term perspective, the anthropocentric one. This is because: (i) the complexity of the biosphere is simplified; (ii) energy bonds accumulated over time produce exergonic reactions; (iii) complex network relationships are dissolved, hence the ecological system requires more external energy; and (iv) self-generation and recycling capacities are reduced to zero. Anthropoplasty exploits the value of the biosphere, squeezes every last drop of it, increases entropy as well as the amount of energy needed, and ultimately makes the system less stable (Odum and Barret 2005). To claim that ecological thinking counters human fulfillment or that it is anti-humanist or against progress is a gross mystification. It is also a mistake to view this conversion as a means of creating a protective sphere, more suitable to human needs. This means confusing a temporary effect with the comprehensive evolution of the system. What we are actually doing is making the planet less and less hospitable to ourselves (Swan and Colino 2020).

In the modern age, *téchne* becomes the strong arm and the accelerator of this process. *Téchne* is not a problem in itself but it becomes one when seen as an instrument of conversion rather than a means of relationship. We could argue that the combination of a triumphalist and celebratory attitude toward the human and the convulsive and rapidly flourishing technological conquests, not yet fully internalized, have accelerated a latent tendency of ours. As far back as the Neolithic revolution, a tendency toward conversion became visible. Although often on the verge of a crisis, our coexistence with nature remained balanced up to a certain point in time. This balance was broken by the advent of the modern age, which can be called the humanist age, because the assumptions of its fifteenth-century forefathers came into full realization. In fact, a consistent trend is visible from the Neolithic Period onward, with sudden accelerations due to incentives such as the discovery of a continent or a technological invention.

The humanist epoch pursues its own coherence through human expansion and the by-products of conversion that represent human power. If the race for transformation erodes the biosphere, it also seals the final victory of the human over nature. Conversion is how the human cries out to the world that it is other than nature, that, rather than immersed in nature, it has completely emancipated itself from it. The humanist era accomplished what the Neolithic period had started: a celebration of work as the sole expression of human value. All around us, desertification is slowly expanding; forests, ecosystems and species—earthly constituents that formed over millions of years—are progressively being annihilated (Kolbert 2015). However, the humanist epoch is not just a Lucullan feast at the expense of the biosphere. It is a project stemming from a precise theoretical framework. We must bear this in mind, lest we risk trivializing the problem as mere hedonism. In the humanist framework the human product is something valuable, while everything else in nature is not. This is what justifies the conversion process.

The Neolithic revolution marks an anthropological change in our relationship with the biosphere that has developed and followed a particular trajectory up to the present day. Just as the Paleolithic period depleted the huntable fauna due to hunting overexploitation, we now have to face challenges the seed of which were sawn in the Neolithic period. If we look at human distribution on the planet and the overall demographic trend, we realize that the Neolithic project of agriculture and animal husbandry is no longer sustainable. A change of pace is necessary, one which goes beyond the exhaustion of resources of the Paleolithic civilization or the conversion of resources of the Neolithic period. Posthumanist philosophy seeks to supplant Neolithic civilizations because the paleo-technologies that exploit and disarticulate natural entities cannot meet the great needs of our planet. We do not have a planet B to terraform, at least not in the short term (Sagan and Druyan 1997). Consequently, the issue of sustainability is not a fad for beautiful souls but a crude necessity.

In fact, the problem emerged during the modern age and its assumption that energy could be extracted from the stocks of coal that millions of years of photosynthesis had produced in the biosphere. The humanist era looks like a time of great illusions, the time when humans believed that their future could be built on the principle of labor. They thought they were autonomous and that *opus* was the only principle worth considering. The project of conversion not only devalued the biosphere, but also dangerously neglected it, focusing exclusively on human hands and human ingenuity. This trend climaxed with the acceleration of the machine age, when technical tools (*tecnai*) were replaced with technology, marking a considerable step toward anthropoplasty (Kelly 1994). From this moment onward, the somatization of the world has seemed unstoppable. If the terms of reference for this project had already been conceived by Leon Battista Alberti, who drew inspiration from the human body's proportions, it is since the invention of hydraulic machines and printing that the process has become increasingly pervasive and has acquired ideological connotations.

This view rests upon one particular characteristic of anthropoplasty: the disjunction of entities and the dissipation of relationships, namely of the system's energy and information content. Anthropocentrism, therefore, is not only a means of bringing the human center stage, but also of considering entities as separate monads with an intrinsic—as opposed to a relational—potential. This intrinsic potential must be converted or enhanced through human labor. As observed, anthropoplasty is a dissipative process, because labor does not produce an increase in value but a depletion of it. Tilled land allows us to obtain edible material; its value, however, is infinitely lower than the forest from which the land was derived. I would like it to be clear that the further we continue with this conversion, the greater our dependence on processes of exploitation will be. We will inexorably come to an end point where the system will collapse by sudden implosion. This is not a catastrophic vision—just a realistic one.

If we fail to realize that this is happening, it is because we are willingly ignoring problems in the hope that someone will come and save us, be it God (for the believers), artificial intelligence (for the atheists) or aliens (for the dreamers). The strong contradiction of humanism is that, on the one hand, it ignores all constraints and fantasizes about a total liberation in the name of a presumed human dignity, on the other, it rejects all responsibilities. It is up to posthumanism to reconnect the human to the biosphere and supplant anthropoplasty—not from a neo-Luddite or antiscientific perspective, though. In the current conditions, it is possible for the posthuman epoch to come about only with the contribution of technoscience. This requires a transition toward the construction of a new alliance whose duration is difficult to quantify. We will need to devote ourselves to relationships and to the preservation of systemic networks by reconciling Prometheus with Epimetheus. Humans have to acknowledge that they are a part of the biosphere, rather than view the biosphere as a threatening or worthless entity to be eradicated or exploited. If we know how to use the connective energy of the biosphere and avoid separation, conscious that breaking down systems in order to dominate them is tantamount to destroying their value, we might well rediscover the feeling of integration.

## *Chapter 3*

# Anthropocentrism and the Ecological Crisis

Carried away by our enthusiasm for the magnificent progressive destiny of the human, we have forgotten about the serious ecological crisis that is looming over our future. This crucial problem derives from our modern-age attitude of considering the biosphere a resource of little importance and at our complete disposal (Burkett 1999). Our neglect is not due to carelessness, however, but is the result of a precise philosophical approach. The anthropocentric model is the great evil of our age: apart from being primarily responsible for our serious ecological crisis, it has caused great suffering to the whole of humanity. Anthropocentrism not only lies behind our devaluation of the biosphere, but it is also the archetype of all forms of segregation and intolerance toward diversity (Boddice 2011). Posthumanism is a systemic philosophy that aims to build a new alliance with the biosphere through the idea of the conviviality of existence. It does not demean the human—instead, it enhances the value of coexistence and the responsibility we carry toward the biosphere.

### NINTH THESIS

Posthumanism rejects the dichotomous and hierarchical classification of entities, admitting a perspective but always integrated plurality, thus rejecting all forms of ontocentrism, regarding any idea of separation as an ontological fallacy.

Anthropocentrism is, firstly, an interpretative model and, secondly, a prescriptive model. It is a generative paradigm. Other than the interaction with the nonhuman, it can influence comprehensively how we interpret the relationship between the subject and the world (Viveiros de Castro 2009). In anthropocentrism, the human stands in the foreground and at the center, separate from the natural. *Essentially, anthropocentrism privileges the observer's viewpoint and his/her needs and expressions, namely ontocentrism*. It should come as no surprise, therefore, that anthropocentrism serves as an archetype for mechanisms of segregation: categories such as the nation, religion, culture, the microcommunity, or the individual stand in the forefront, above and in opposition to dialectical otherness. Anthropocentrism becomes not only an anthropological but also an identity-based machine (Agamben 2004). As well as "teaching" us who we are, this machine also creates two levels, one above the other. The problem is twofold: (i) it is not true that identity derives from opposition, because all inheritances are shared and each predicate is always a relational emergence; and (ii) it is impossible to correct the mechanisms of exclusion and discrimination by only tackling their resulting forms (e.g., racism, male chauvinism, etc.) and not their generating center, because it will keep on creating forms of segregation.

Posthumanism views anthropocentrism as the most natural, immediate and, I dare say, easily justifiable expression of ontocentrism. Anthropocentrism is also the core of the human problem. It is a point of vulnerability that must be tackled critically because it generates problems that have always persecuted humanity. Anthropocentrism encourages ontocentrism, a feeling of importance and superiority that prevails whenever otherness is transformed, more or less explicitly, into a beast to be subjugated. The process of subjugation at the heart of ontocentrism is linked to the anthropoplastic conversion. Subjugation is a defense mechanism triggering a need for control (Deleuze 1994) and predictability that distorts all forms of diversity. Ontocentrism is an attitude that opposes polycentrism (Descola 2016), even though this dualism ultimately fails to grasp the paradigmatic revolution operated by posthumanism. Posthumanism does not seek to legitimize a plurality of ontocentrisms, but to question the principle whereby subjectivity can only emerge from a central and disjointed positionality.

As already mentioned, posthumanism does not mean being above the human, nor is it a form of anti-humanism. Yet it cannot be considered a form of neohumanism, either. The prefix "post" urges us to go further; it has the same function as a transitional object (Hayles 1999); it is an appeal to overcome the weaknesses that make the human dangerous to the biosphere and to itself. First and foremost, posthumanism rejects the notion of anthropocene (Crutzen 2005). In a legitimist or critical sense, the term "Anthropocene" underlines the fact that the human pervades and characterizes this epoch; however impactful and invasive, our species is yet only just one of the elements of a network that contains many connected and connecting factors. The notion of anthropocene conceals, once more, the ontocentric principle. This is suggested by: (i) the idea that the human may be the cause or the control center of what happens in the biosphere; (ii) the disjunctive perspective that considers the anthropic the outcome of a process of extraction, like an alchemic process that filters the purity of a substance; and (iii) the hypothesis that the human dominion over reality is the result of progress rather than of greater dependence on reality (Stiegler 2013). A term like Anthropocene—which I believe is wrong, because it is the fruit of human myopia, a tendency to emphasize human presence—risks endorsing an ontocentric perspective.

The Anthropocene could also be interpreted as an appeal to take critical responsibility for the future of our planet. The concept of Anthropocene itself, however, assumes that the human has gained so much power that it may affect and define a biological epoch, which is not the case. The Anthropocene is just another product of the segregating machine previously described: it rejects the principle of sharing and views predicates as essences and as self-referential rather than emerging from relationships (Bateson 1976). The Anthropocene is an oneiric by-product of modern anthropocentrism, an expression of the human celebration that was sparked off by the technoscientific revolution of the so-called short century. Anthropocentrism has transformed sci-fi production into a transreal condition wavering between dystopia and mirage, where the depiction of the future is actually a representation of the present. Transreality helps us understand the real through metaphor; it talks about the future in order to shed light on the present (Rucker 1983): in the film *Blade Runner*, for example, the exploitation of the replicants recalls forms of social exploitation that occur in the contemporary world. This takes us back to ideological anthropocentrism, a triumphalist *leitmotif* rather than a mortgage for the future, a dualism that marks the shift from humanism to posthumanism. The project of human celebration that has developed among modern authors is not something new, nor did it begin after the Middle Ages. It was during this time, however, that the human gained a new self-awareness and converted it into a programmatic platform. Since its fifteenth-century forefathers, anthropocentric assumptions have become a paradigm and a project consolidated through technopoiesis.

Anthropocentrism has been described as the almost spontaneous product of an entity that comes into the world and becomes stronger through repeated external confirmations. The machines we have produced offer an external support that makes us believe that we are strong and powerful. By increasing control, technoscientific development fuels self-referentiality. However, it was fear that mainly helped build up the walls of centripetal subjectivity. Fear has a profound significance in human ethology (Eibl-Eibesfeldt 1997). The human species has scarce predatory skills—it is a generalist and an omnivorous scavenger, but above all prey, grown up in a savannah teeming with threatening species. The human is highly sensitive to the lure of fear, which means that it responds not only to actual external emotional stimuli, but also to endogenous and more general elements that activate the alarm system. The ideology of fear (Bourke 2005), also present in various postclassical sources, strengthens centripetal points of reference and fuels the segregating machine: it transforms the fear of being attacked into a justification for aggression; the aspiration to self-determination into colonialism. The ontocentrism concealed in anthropocentrism—its most impulsive and banal expression—may lead to innumerable forms of discrimination. To focus on specific forms of segregation and to believe that, by amending them, we can solve the issue of discrimination altogether means distorting the problems and taking our attention away from them. Actually, the mere promotion of such amendments risks creating new forms of exclusion.

Fear is an emotion that is useful for survival (Damasio 1994). While it has phylogenetic reasons to exist, if it is magnified, it produces behaviors that may cause vulnerability in humans. Let us go back to the need to have a critical trust in human dispositions. The ideology of fear, even in its heuristic function as argued by Hans Jonas (Jonas 1984), risks worsening a series of ontocentric attitudes, among others: reducing an individual's critical skills, enhancing our tendency to follow autocratic directives or directives dictated by the mainstream, emphasizing disjunctive polarizations and encouraging aggressive tendencies. An excess of fear or a condition of permanent fear increases our tendency to create walls, mistrust others, look for scapegoats (Girard 1994) and blindly follow centripetal points of reference. I would like to argue that fear is the real foundation of ontocentrism; it is what marks the distance between what we consider a pure identity and what we view as otherness. The posthuman era therefore cannot rely on fear, not even as a heuristic condition. It must rely on the liberation from fear and the emancipation of humanity from a condition of permanent panic. It will also need to be built on the principles of self-donation and hospitality (Marion 2018) and avoid all forms of segregation.

From where shall we start, then? Because of its implications, I believe the starting point should be a new alliance with the biosphere that may overcome anthropocentrism. If we do not deal with the problem of anthropocentrism head-on, we will never be able to face its ghost, ontocentrism. This is an epistemological mechanism that assumes that an entity can be extrapolated through a process of purification that makes it self-referential and separate from other entities (Stiegler 1994). If anthropocentrism is the archetype of all expressions of discrimination, the humanist mythopoiesis has undoubtedly magnified its problematic potential.

Anthropocentrism is a process of hierarchization that may follow different routes by appealing to different predicates, such as reason or language, the ethical or political dimension or the capacity to question or believe. Even the theory of human incompleteness (Gehlen 1987), according to which nature was like a stepmother who refused to bestow on humans the equipment necessary to adaptation, ultimately becomes a dialectical strategy that produces a triumphalist synthesis. Anthropocentrism mainly rests upon tautology, circularity, and the assumption of an overlap between the human and anthropocentralization. The latter marks the breaking point with posthumanism, which, vice versa, views the human as an expression of *anthropo-decentralization* (Marchesini 2002). Therefore, we should not think that posthumanism simply views the human as the expression of the hybridizing tendency of human nature rather than viewing it as the expression of an essence. Shifting from an ontocentric to an ontorelational perspective has profound repercussions for some—surely the most important—critical issues related to the subject's lack of existential meaning, convivial communion, and critical skills. We need an epoch that, rather than repudiating finitude or the organic body as an expression of palaeontology, repudiates war and discrimination.

## TENTH THESIS

Posthumanism is not an asystemic disruptive philosophy, but posits a different systemic approach to humanism that is no longer centered on the orbit around the autopoietic essence, but based on symbiontic participation.

To understand posthumanism, it is necessary to shift from autopoiesis to sympoiesis (Haraway 2008) in three fundamental ontological points of reference: (i) inherence or the entity's phylogenetic legacy; (ii) development or its process of emergence; and (iii) expression or its manifestation. In an autopoietic perspective, the entity is also considered interactive. However, this happens within an essentialist framework, which considers the genome an inherence, embryogenesis the formation of individuality, and predicative emanation the expressiveness of the entity. On the contrary, in a sympoietic perspective: (i) inherence means a shared phylogenetic path, shared predicates, introjected relationships (*Having-been-there-before,* Marchesini 2020) and a heteronomic predisposition; (ii) development means "system of development" and not embryogenetic individuality, because the organism's biological unity has been co-constructed; it is the co-factorial product of sundry epigenetic and ecological entities; and (iii) expression refers to the emergence moment the relationship produces: all manifestations of the entity are never autarkic so they cannot be inferred by studying the entity from within.

The posthumanist proposal should not be confused with deconstruction (Derrida 1976), nor with a lack of predicative synthesis. On the one side, there is an essentialist ontology whose architectures do not affect the coming into being of the entity, but only the system of dependencies, collaborations, competitions, etc.; on the other side, there is *symbiontic ontology*, whose three fundamental ontological points of reference are sympoietic products. The concept of symbiosis (Margulis 1981) shifts from ecology to philosophy. It reveals that we cannot view the entity outside of the relationships that have structured its inherence, development and expression. From a philosophical viewpoint, posthumanism has problematized some of the scientific evidence of *evo-devo* theories (Gilbert and Epel 2015). They take account of a morphopoietic theory of form and a life phenomenology that clearly depart from essentialist ontology (Barbaras 2022). As previously observed, in its original formulation, even Darwinism undermined the essentialist view of the entity, via populational nominalism, the sharing of predicates between entities and the reversal of the meaning of mutation—previously considered a monstrosity, since Darwin mutation has been viewed as a source of evolution. In the posthumanist approach, we need to speak of a *Darwinian philosophical laboratory* that has been profoundly changing life phenomenology through the contributions of various scholars (Sterelny 2007), all fundamental to step away from essentialism.

There are two points we need to take into account. Firstly, the Dionysian concept of the river of life (Nietzsche 2012) that considers entities part of a sole great source. That species originate from a single source implies that they share traits, and that each species has attributes deriving from selective processes preceding its origin (e.g., we have four limbs, because the first tetrapod was so selected). By previous existentiality, memories that have not been experienced yet affect the subject's openness to the world, I obviously do not mean metempsychosis. What I allude to are existential sediments that had their significance along the river of life, across the trans-specific *continuum*. The *having-been-there-before* is not the recollection of someone who preceded me. It is the pandemonium of memories, mostly nonhuman ones, that reveals how the very inherence which makes me open up to the world is something that precedes my being human (Scheler 1961). The *Having-been-there-before* is the opposite of Heideggerian *Thrownness*: it combines wonder with the joy of rediscovery. Our remembering is blurred not because of amnesia, but because it is unidentified and choral. The humanist solitude of Being is the opposite of the *convivium* of posthumanist philosophy: life is participation, sharing and figuration, namely images and representations defining themselves across time. To be alive means to recap and add to the great chorus of existence. Etymologically, this image suggests a collective gathering around a center. This center, however, no longer represents a superior being. The chorus ultimately suggests the concert of all that has preceded a particular form of existence, and gift as an expression of sense.

The second point we need to take into account is development, acquiring a form. This no longer means, as in the concealed essentialism of twentieth-century genocentrism, translating a recipe (Dawkins 1976). It rather means opening a construction site that envisages external input but also suffers from the inevitable variability due to chance, which may repeatedly influence the entity's phylogenetic trajectory. The core concept is, once more, symbiosis. Becoming human means opening up to the world through an ontopoiesis that is neither predetermined nor amorphous. Ontogenetic virtuality (Gilbert 2000) defines the boundaries of the ontopoietic construction site, just as environmental characteristics define the opportunities and limits of an eco-systemic emergence (Oyama 2000). However, as in nature, the niche is not an empty space to be occupied (Odling-Smee et al. 2017), but the result of the co-constructive interaction between a species and the outside world; an organism acquires form by means of various genetic, epigenetic, and symbiontic actors. For example, today we know that two of our most important interface systems—the immune and neurobiological systems—are strongly influenced by symbiontic bacterial populations (Gavin 2018). Yet the genome also contains proviral elements, and the epigenetic shell hosts populations of endosymbionts.

So how can we possibly continue to refer to ontology by traditional parameters, such as "essence" and "the individual"? May we confine scientific evidence to a biological description without affecting philosophical analysis? It is not a matter of biological reductionism, of reducing problematization merely to description. We need to ask new questions on the basis of our new findings, because there should be no divisions or barriers between the different domains of knowledge. The co-constructive view of ontopoiesis (Lewontin 1983) requires us to change our interpretative points of reference about what becoming organism means. Though morphologically a specific unity, the organism cannot be considered autopoietic, authentic, and pure. Heteronomy not only means that an entity needs external contribution to complete itself. It also means that multiple morphopoietic directions are possible, depending on the ecological parameters present (polyfemism). Hence the term "conjugation" used to define non-accidental morphopoietic plurality. An example is provided by the sex of turtles: it is acquired during embryogenesis and is dependent on the thermal gradient (Crain 1998). Therefore, we need to revise how we consider the relationship between the organism and technology. We must go beyond the humanist approach that tends to view tools as either external additions—a garment—or an empowerment of human traits orthe human's inherent predicates. We must speak, that is, of a techno-mediated *evo-devo*.

Another aspect that requires consideration is the concept of development. Development refers to a permanently ongoing ontopoiesis, rather than to the pursuit of a preestablished existential objective. This marks a shift in perspective that transforms development into a never-ending itinerary without a goal. The lack of conclusion and purpose converge into the concept underpinning the posthumanist proposal: that of mutant identity (Alfano Miglietti 2008). Each of the above points, however, focuses on a particular aspect of ontopoiesis. As development progresses, inherent virtuality is depleted. The gradients of dispersal depend on the situation, yet virtuality will never be entirely consumed. Dispersal is constantly occurring, because the subject's ontopoietic construction site remains open to the symbionts' intervention. An individual's path is never linear, coherent and stable: its characteristics can always be identified only by an arbitrary and *ex-post* process, by connecting dots that might have been scattered in infinite different ways. The journey of life is always an adventure into the unknown, an improvisation, it is dialogical creativity rather than a project pursued with coherent determination.

The symbiontic view of the entity hardly makes us guardians of the *Being*. It makes us anything but the helmsmen of a ship heading for a safe harbor. Herein lies, on the one hand, the deconstruction of ontocentrism, and, on the other, the focus on the eco-ontological character of being, which has no hierarchical points of reference. The child is not a function of the adult, nor the larva of the imaginal form, nor the project of the result, and desiring is not the lack of what is desired. No child grows up to become an adult, but lives its life in full, age by age, day by day, reshaping its ontological castle on the basis of circumstances, such asthose symbiontic friends who keep it company along its life path. There are no failed projects in life; only stories, each with its dignity and contents, which get integrated through processes of relationship (Noble 2008). Symbiosis refers to the condition of conviviality whereby an entity forms and expresses itself, because, if entities were separate, they would have no possibility of predication. Symbiosis therefore represents the posthumanist systemic. Symbiosis is the guiding principle that helps us understand ontopoiesis in all its facets as well as the fact that we are always anyhow connected to otherness, in both space and time. This is why we need to improve our capacity to read symbiosis and we must refuse the logic that considers entities only conflictual and competitive (Dupré and O'Malley 2009). Cofactors are present and constitutive of conflict too.

We would not be able to understand the principle of symbiosis to the full, however, if we clung to an essentialist ontology, imprisoned by the idea that the entity is self-defined and separate from all others. Indeed, before defining where the center is, we need to discriminate and separate entities, and view them as independent. Separation arguably represents the premise of existential solipsism. Posthumanism, in its different expressions, tends to liquefy crystallized entities. It prefers a fluid, morphopoietic dimension, which is also transitory and metamorphic. This is what makes the posthumanist proposal revolutionary: it rejects pre-constituted models, particularly when they are portrayed as absolute truths. While refusing relativism, posthumanism observes the different expressions of reality in their plurality, even prior to considering different viewpoints. This is because the same phenomenon has always had different levels (Nietzsche 2020). Each of these levels tells us about a precise relational systemic that calls upon multiple actors of onto-poietic symbiosis.

## ELEVENTH THESIS

The biosphere is not a stage or a home for humans, nor is it a stockpile of available resources, but an organism comprising the interweaving of different biological relationships maintained by homeodynamic network mechanisms.

The future of the human undoubtedly depends on a radical change in our relationship with the biosphere. We urgently need an out-and-out conversion that helps us come to terms with a number of problems which will otherwise not only worsen, but converge in a sinister synergy. Terms such as "radical reform" and "spiritual conversion" are not inappropriate. The ecological collapse is not simply forcing us to face operational problems, but value structures. I have already touched upon this issue when discussing the anthropoplastic principle of the humanist tradition: because work was perceived as enrichment rather than dissipation (Patel 2009), transforming the biosphere into a human product was tantamount to increasing its value. A second important point to discuss is how the biosphere is viewed as a stockpile or collection of entities: the space they occupy can become available for human use, while our annoying roommates turn into useful and exploitable resources. We lack a properly ecological reading of the biosphere as an organism of which we are part, fruit of an interweaving of biological relationships. It is no surprise, therefore, if the crisis is ecological and not only environmental (Naess 2016).

If we find ourselves in extremely critical conditions today, it is because the human integration in the biosphere was badly set up, starting from the view of our position in nature or in ecological dynamics. "Where did we go wrong?" is the first question to ask. This is propaedeutic to setting up a change. We must avoid falling into the trap of blaming single dysfunctional strategies and thus miss the core of the problem (Klein 2015). To adopt an ecological perspective means to reject the interpretative keys with which we stretched and distorted our explanations of the biosphere thus creating impactful operational models. An irreconcilable incongruity separates *eco-logical thought* from the humanist approach, which cannot comprehend the integrated and inclusive dimension of life phenomenology. The humanist framework envisages an unlimited exploitation of the biosphere, or at best, only research into less impactful practices that may increase the space available in terms of either time or extension. By so doing, we lapse into oleographic conservationism, as if the environment were something fixed and steady. This prevents us from understanding the massive connectivity of the biosphere and the dynamic and metamorphic meaning of life.

In the humanist model of the garden (Thomas 1983), the biosphere does not stand for generative complexity but represents an unbearable mess that needs tidying up. The garden only acquires meaning when inhabited by the human and profiting from maintenance work. Deforestation does not only imply recovering biomass and having land to till. Deforestation equals lighting up a dark ravine that harbors threatening entities. The humanist sees the biosphere only as a backdrop (Midgley 1981), as a house or an aggregate of living beings to which he or she does not belong. The Earth becomes a place of transit or a workshop, which helps us reach another existential dimension. Earthly life is the price we have to pay to earn the ultraworld. Since the work of Pico della Mirandola, the thrilling pursuit of the human ontological ascent has not only led us to claim superiority over other living beings. It has become a process of emancipation that shakes off the telluric in order to gain a new, no longer natural, dimension. Something similar happens with the transhumanist desire for immortality (Bostrom 2016): albeit immanent, it implies an emancipation from nature. The Earth, whereby we also mean the natural condition, becomes a prison of the soul, a valley of tears and an "opaque atom of evil" (Pascoli 2019). It is impossible to comprehend today's ecological crisis if we continue to ignore the persistent campaign of devaluation that the humanists conduct against nature.

The humanist perspective transforms earthly life into a Dance of Death (Tenenti 1989), a *Being-for-death* (Heidegger 2001) and an ephemeral transit that negates life. This imagery mainly derives from the classical Platonic and Stoic tradition, with its emphasis on the individual, and from the Abrahamic tradition, which is both naturophobic and rooted in the idea of original sin. The most striking aspect of this vision is the subject's solitude, its existential isolation (Schopenhauer 1884). Withdrawn from conviviality, the subject has no shared belonging. In more general terms, the subject does not feel it is part of the biosphere as an organic whole. This loneliness, an existential *divide et impera*, locks up the subject in fear. Ashamed of its nature, the subject needs to find scapegoats to rediscover moments of brotherhood. The individual finds in the thrill of verticality a way to sublimate impulses by wrongly assuming that they can thus be discharged. In fact, these impulses turn into deviations, making the subject more and more dissatisfied and prey to neurotic languor (Esposito 2011). While humanism is undoubtedly the ideology of the conversion, through labor, of the natural into the human product, our compulsive approach to work, nurtured by loneliness and dissatisfaction, fuels our continuous need for self-assertion.

Loneliness condemns the individual to a feeling of existential vanity, an unbridgeable gap that can only be filled with existential artifices (Taylor 1991). Hedonism, the yearning for experience and excess and the need for self-affirmation, can be considered natural only in part. It is because the mechanisms of self-affirmation have been adulterated that these needs and expressions have become compulsive. Loneliness triggers a desperate search for existential meaning in a cultural climate that induces the individual to revert to itself. This leads us to conceive of existence as consumption (Baudrillard 1998), as a need to burn intensely so that we can say that we have lived. This centripetal existential rotation explains why we cannot understand what ecological conviviality means, what that *Being-with* is, which reveals the ontopoietic error of separation. Finding the meaning of existence within one's life—biography—or in individual life—as a separate atomic entity—is impossible. Humanist loneliness makes the human the slave of its own fears, always dissatisfied with its search for meaning.

The biosphere, despised and neglected, becomes the setting of a tragic play, barely suitable for the monologue of an actor craving for success. Spotlights illuminate the actor's excessive gesticulations and profusion of words, expressions of his vain need to be acclaimed. All other presences are erased, and the landscape is transformed into a two-dimensional backdrop. While the actor is waiting for, or dreaming of, applause that is slow to come, and which could hardly fill his void anyway, no panorama is visible, only a *trompe-l'oeil* effect. Perhaps the theater is empty and the actor cannot be sure of anything, except his words. So he believes that the only sense he can make of his spectacle is to bring it to a close as early as possible. In the meantime, however, he has come to hate his audience, pity his own misery and, most of all, resent the theater, peripheral, dilapidated, limited and beneath his dignity (Accarino 2015). The theater is a metaphor for the biosphere. Humanism has thrown us into a deadly loneliness that seemingly encourages our industriousness. Industriousness is actually a compulsive effect caused by the anxiety of our everyday lives. The ecological problem is not primarily about the human dominion over the biosphere. The craving for dominance and control is also the fruit of loneliness, of the hatred of nature that underpins humanism, or a convergence of anthropocentric traditions.

What this imagery implies is the devaluation of nature and the biosphere on the one hand, and, on the other, the existential loneliness of the individual. As long as it persists, we will not be able to appreciate the extent of the ecological crisis. We will view it only as an environmental urgency and a resource deficit, as suggested by the widespread metaphor of the Earth as our burning home (Shankar Jha 1998), or by the news of energy shortages or famines we get from the media. By clinging to this perspective, we can conceive only of minor adjustments, such as the green economy or waste reduction. However important, these recommendations will not solve the problem: the gap between the aggravating factors and the interventions implemented will inevitably tend to widen. When I argue that the posthuman era is a period of transition, I mean to underline the urgent need for a profound and paradigmatic cultural change (Easterling 2021). This process will inevitably require a lengthy period. But it will also have to tackle very serious problems in a very short time. The preparation of this paradigmatic metamorphosis will require commitment to operational changes in the short and medium term. What must be clear, though, is that these changes in the structures of societies will affect the lifestyles of millions of people, and the economic organization of international production and finance will have a high impact and cannot be conceived as an eco-friendly makeup.

The devastating impact of these measures will most likely encourage that paradigmatic shift, without which we would not even approach the root of the problem (Foer 2019). The biosphere is not our home; in fact, we are only just a tiny tissue of this organism. The loneliness of the subject melts like snow in the sun when we recognize our anastomotic continuity—and our temporal continuity—with all living beings. By continuing to address the question in terms of inhabitability instead, we will stay in the problem, namely cling to a functional view of the biosphere as extension. The biosphere is depicted in spatial terms because what still prevails is a view of nature as extension (Kauffman 2019). We keep on hearing that we ought to protect and preserve our home. However, it is the very notion of home that prevents us from appreciating that we are ontologically entangled with the biosphere. If we cease to view the biosphere as an environment—a home or a stage—and adopt an organismic and inclusive view instead, we will discover, firstly, that the subject is not lonely, because individuality is only a façade, and, secondly, that our body expands and is, therefore, affected by our actions.

The very principle that informs our search for meaning has to change. We should redirect our focus: from the individual to its relational thresholds, namely the process of integration. We exist because we are a function of the biosphere and not vice versa. The current crisis was ignited by the humanist concept of the availability of resources and worsened by the insatiability of human nature and the possibilities offered by technoscience. Since the values and moral criteria of humanism have caused the crisis, it would be foolish to imagine that they can also provide strategies to solve it. The environmental movement of the past century made no impact, nor did it offer useful and reliable directions (Passmore 1974). It was far too immersed in conservationist and disjunctive mirages about nature and the human dimension to implement the principle of integration.

## TWELFTH THESIS

Posthumanism recognizes the serious ecological crisis we are experiencing as the most important problem of our time, regarding it as an issue that directly affects the biosphere and thus the human, and which calls for a eu-topian project.

There is no doubt that the posthumanist movement has been repeatedly misinterpreted. It has been considered a project about the overcoming of the human as a species, translated into a sci-fi imagery of cyborgs and artificial intelligences moving against a backdrop of technologies devoid of natural elements (Tortoreto 2018). The future is thus portrayed as a time when, rather than interacting with natural otherness, an empowered human only interacts with its technological interlocutors, the machines. Since machines are human products, however, the human finds itself in a mirror maze, which multiplies and deforms its own image. In the transhumanist imagery, nature is totally absent. The only admitted interlocuters—albeit only potential ones—are artificial intelligences and alien entities. In this totally denaturalized existential dimension, even the human sheds its organic *exuviae*, the body, the anachronistic legacy of a past from which the human is eager to take leave (Kurzweil 2005). If this condition did not appear to be, first of all, ridiculous, I would consider it tragic. It betrays an enduring state of humanistic hypnosis brought to its extreme consequences. It is a tragic condition because it prevents us from realizing that the real challenge of the posthuman era is not how we will succeed in floating in the technological ether; it is how we will survive the looming ecological storm of which we will be the first sacrificial victims.

Anthropocentrism is the actual problem we need to tackle. Its significance no longer concerns the ethical domain only, as it did in the past, when anthropocentrism was not considered in its multifacetedness but was limited to the anti-speciesism debate (Singer 1975). What needs revising is first of all our ontological anthropocentrism, namely, autarky, universality, centripetation and disjunction or separation—all aspects deriving from the humanist matrix. Posthumanist philosophy is a critique of the humanist paradigm. On the one hand, it goes beyond a conception of the human *iuxta propria principia*; on the other, it continues the process of human liberation initiated by our humanist forefathers, albeit no longer from the viewpoint of an emancipation *from* nature, but of an emancipation *of* nature. Posthumanism means rediscovering an integration with the biosphere, celebrating conviviality with and participation in the living world. Going beyond the Cartesian concept of nature as extension was the starting point. After the humanist intoxication, posthumanism should be like an *awakening of consciousness* (Baioni, Cuadrado Payeras, and Macelloni 2021): it gives us the opportunity to put center stage and problematize our relationship with the biosphere. Rather than being a separate entity coming to terms with us as interlocutors, the biosphere encompasses us. We are all members of an assembly striving to overcome disjunctive obstacles.

Posthumanism is therefore the ability to ask new questions, questions ignored by modernity, because the principle of extension was supposed to provide all the answers. The posthumanist proposal focuses on the paradigmatic aspect of humanism, giving top priority to eco-ontological integration in its programmatic agenda for the near future (Verducci 2008). This is an existential dimension that could be baptized *Being-for-life*. Endorsed by a philosophy of the relationship with the nonhuman universe, it envisages human participation and involvement in the biosphere, with a view to inspiring a *new sentimental education* (Marchesini 2022). Posthumanist imaginary will thus be able to include the technological dimension, but in a totally different perspective from the hyperhumanist one and its transhumanization into a postorganic condition disconnected from nature. In posthuman imaginary, the ecological crisis is paramount. Not in dystopian or utopian terms, though. Posthumanism strives to single out the best forms of integration with the biosphere by considering correlative and convivial factors. It views the dialogue with otherness as an indispensable starting point to prevent relapsing into anthropocentric mirages. Even from an imaginative viewpoint, this representation is very different from the technological kaleidoscope. It is a *Eutopia*.

Eutopia means looking for the best possible form of convivial integration and formulating a project of coexistence that will contribute to solving the ecological crisis. It starts from a scientific assessment of the elements at stake. Eutopia is the opposite of dystopia: with its feet firm on the ground, it believes that, regardless of its limitations or, actually, because of them, the natural condition is the existential dimension most suitable for the human. Unlike utopia, which alludes to a non-place, disregards existing conditions, and is the fruit of an anthropoplastic ideology, Eutopia is an ecological model rooted in a scientific assessment of the agents involved. With its solipsistic and emancipatory human vision, humanism has disseminated a large number of utopias that turned out to be a source of suffering for humans as well. Posthumanism aims to supplant the concept of utopia once and for all by drawing inspiration from ecological models and applying their laws (Odum and Barrett 2005). The most explicit difference between transhumanist and posthumanist imagery consists of this critical view of reality, which is hardly prone to the dreams and reveries of *Homo faber*.

However, Eutopia should not be confused with a holographic image of a wise and harmonious nature that needs preservation and protection; with the technophobic neo-Luddism of twentieth-century environmentalism; with the syncretistic and new age vision of a provident Mother Nature; with the Leibnizian panglossianism of the best of all possible worlds or with an animistic anthropomorphism that is the outcome of our own projection (Gould and Lewontin 1979). For posthumanism, Eutopia is not merely a return to nature, but a horizon to conquer, a project to realize through, for example: (i) a disposition to recognize otherness, both as something we integrate within ourselves and as ontic plurality; (ii) the promotion of scientific knowledge and the development of neo-technologies; (iii) the development of new social structures and projects of citizenship; (iv) advancements in agriculture and animal husbandry that may replace the Neolithic model; and (v) research in new processes of energy use. This model does not contemplate the possibility of an emergency lifeboat (Hardin 1974)—should the human ever plan to abandon the Earth to its fate—nor a transformation of the world into an amnio-technological sphere. This is why the ecological crisis is the necessary starting point for any reflection about the future of humanity.

This is not about generating fear. It is a lucid assessment of the serious risks we are facing. Critical issues include: (i) climate change and altered weather conditions in various geographical areas; (ii) alteration of the most important homeostatic cycles, such as those of nitrogen, carbon, water and acidification; (iii) the destruction of biodiversity and of the integration networks of living organisms, with increasingly significant epidemiological consequences; (iv) the drastic depletion of energy resources, deforestation and the decrease of land for agro-food processing; (v) the increase in concreting, environmental pollutants, groundwater and soil pollution, and plastic in the oceans; (vi) the exponential growth of human populations, a parallel increase in energy consumption lifestyles and in the resulting impact on the biosphere. This is not to draw a catastrophic picture that would drag us back to the condition of removal typical of dystopia, but to show how, on the contrary, our existential points of reference are profoundly linked to what, with guilty ease, we are jeopardizing. Our connection with the biosphere is much stronger than we are willing to admit.

By analyzing the aetiology of the various ecological problems, we realize that there are many causes and they need to be dealt with in their complexity. It would be wrong to relate all ecological problems to a single factor. If we observe the behavior of living organisms, we notice that both the reproductive drive and the invasive tendency are the fundamentals of all species. As responses to competition and attacks, they are rewarded by natural selection.

They are later tempered by the counteroffensive of other species. A good balance between the two forces at stake is thus created, a dynamic balance that is usually structured as a network of relationships between species (Quammen 2018). Competition also strengthens correlation, namely that set of qualities that we call adaptive, but which, more correctly, should be called correlative, because they refer to external selectors. The human has undoubtedly demonstrated great affirmative abilities since the Paleolithic Age. Through the development of techniques and highly performative technology, our species has avoided mechanisms of ecological containment.

In a logic of integration, it is the human who should find solutions to the problems it has caused and avoid using the potential of technology to justify its arrogance. It is no longer time to eulogize about human greatness or to tribute a *triumphus* by displaying the rich spoils (*spolia opima*) of defeated nature, like the Romans used to do with the corpse of the enemy general killed in battle. We must evaluate the aspects of vulnerability inherent in this attitude. After all, this is what we dispute in the humanist vision, which is still ensconced in an imagery that might have made sense for the early humanists. Because it ignores the ecological crisis, however, today it reveals a disarming myopia. The memento that should hover over our actions is that time is running out and that our chances of preserving this wonderful world diminish with each passing day. As we speak, extraordinary natural spectacles are being erased, daily, because of our negligence, because of that distraction that makes us run after technological toys, trivial virtual worlds, and toxic hypnoses that curb our opportunities to experience happiness.

While there are a thousand reasons that may have pushed us to the brink of this ecological crisis, our blindness toward its consequences is no doubt due to cultural limitations: we prioritize human profit, even the most trivial one, over ecological competence. Micro-dimension of planning and management of green areas in the city, the way agro-livestock production is set up, the negligence with which we manage the few natural spaces left, and scarce attention to ecological education are revealing. We have cultural rather than operative limitations: most of the professionals who should be working on how to (re)connect the human to the biosphere only have humanities backgrounds. Rather than rambling on about the urgency of drastic measures in the vain and childish hope that someone will come and save us, it is our incapacity to give this serious problem due consideration that requires attention. Ecological issues stand out as one of the most relevant aspects to tackle in the transition period between the humanist and posthumanist era. We need to not only take measures to deal with the issue of energy sustainability and anthropic impact, but also address the paradigmatic core of how we conceive our relationship with the biosphere.

*Chapter 4*

# Building New Experiences

This chapter represents a crucial moment in the discussion this manifesto hopes to elicit. It deals with the individual's experience in an era of transition, when some of our certainties and points of reference are collapsing while others have not yet fully appeared on the horizon. It is very important to bear in mind that posthumanism does not mean anti-humanism: human dignity is not in question. What comes into question is how we view human dignity. Posthumanist philosophy differs from transhumanism because it brings the subject's somatic condition and the ontological consequences of *Being-a-body* center stage (Merleau-Ponty 2003). It is not an impermeable or a withdrawn body. And it is not a perfect or a deficient body, either. It is actually a welcoming and hospitable body. The corporeal dimension is like a matrix of relationships that includes otherness (Scheler 1970). We speak of an ever-changing somatic plasticity and thresholds of hospitality events (Stein 1964). This chapter deals with the consequences of becoming corporeality. It is a warning against solipsistic and narcissistic drifts as well as against anti-humanist ones. Freeing the body does not mean falling into hedonism, but feeling part of something that transcends the individual by admitting our need for connection and sharing.

## THIRTEENTH THESIS

Posthumanist philosophy proposes an existential revolution in the relationship with the otherness of the biosphere and the technosphere, starting from a reappraisal of the somatic dimensionality of the human or the liberation of the body.

The somatic experience, namely how to conceive the bodily dimension—feelings, desires and expressions—in a perspective of liberation of the body, is one of the core aspects of the posthumanist conversion. Passions are usually believed to be forms of imprisonment. Posthumanism goes in the opposite direction and shows that passions liberate us. What I refer to, however, is a very different liberation from that of the second half of the twentieth century. Indeed, back then, the liberation of the body was only an illusion. It actually increased the captivity of the subject (Jaeggi 2014). Practices of somato-addiction, such as the craving for objects, erotic hedonism, the ostentation of status symbols, food adulteration, overconsumption of animal foods, and addiction to psychotropic substances (Benasayag and Schmit 2003) only made the individual more and more segregated. What was passed off as liberating only added to a feeling of permanent dissatisfaction. Now that we have experienced firsthand that happiness or, at least, levity, does not come from having or being, whenever having or being become forms of self-assertion, it is possible to start a discussion. We only need to watch and read commercials to realize that we are prey to a huge deception. Advertising creates toxic feelings of lack and inadequacy (Benjamine 1921); it makes us gloat and ensnares us in the illusion of being in full control of our choices and life plans.

We claim we have become emancipated simply because we are living in a world where coercions have been transformed into hidden persuaders and our bodily needs are being fed by addictive surrogates. They are addictive because they are not fulfilling. They provide us with only just enough satisfaction to keep our addictions alive and hold us in a state of permanent languor (Greenfield 2015). A society in captivity is the founding principle of capitalism, the mightiest hindrance to building a convivial society (Patocka 2016). It is addictive because it keeps the individual in the prison of narcissism and denies mutual dependence, which is the founding principle of dialogue. Freeing the body experience means eliminating all forms of captivity. This is not in order to achieve a mystical condition, though, which rejects somatic phenomenology anyway; it is to make room for the body by achieving a harmonious balance between desire and fulfillment that erases the need for surrogates. Liberation of the body means emancipation from all forms of egocentric imprisonment. We can also achieve this emancipation by getting to know our body's characteristics better, including what predispositions make us vulnerable to ideological deceptions. We should not deny our passions but be aware that they can be lured by dangerous sophistications, first of all, by denying our animal condition. We need to break the chains of this drug-addicted life.

Even though it began with legitimate demands, the 1968 revolution of customs increased our dependencies. We became dependent, firstly, on the thousand plateaus of the generational market (Anders 1956), and, secondly, on anxious yearnings for self-fulfillment. We began to search for ways in which we could confer meaning and value to ourselves. The imperative of biographical fulfillment—no matter whether hedonistic, experiential, professional or power-related—produces an existential languor that will remain forever unsatisfied, simply because it is ontologically wrong. This orbital movement around the pivot of the Self increases an individual's yearnings and dissatisfaction. The individual is inextricably tied to the sources of dependence that imprison it. The subject's gravitation around itself and the externalization of pleasure, namely mistaking the desire to do (i.e., express one's motivation) for the possession of an object (i.e., the target of our doing) has certainly not contributed to liberating individual corporeity; instead, it has only transformed the individual into its own captor (Benasayag 2015). The body is apparently pampered, and the individual feels he/she has been liberated, but it is an illusion. Post–World War II generations are dearly paying for this mistake, which also lies at the root of the current ecological crisis.

The exasperated individualism of the twentieth century is but the latest phase of a trajectory that has boosted disjunction and separation (Lasch 1979). The roots of these processes, which characterize the modern age, lie in humanism. Compared with an individualist living in a capitalist society, the convivial human has less need to assert himself or herself to feel alive. Feeling part of a group makes him/her more moderate and less narcissistic (Levinas 1998) even in his/her self-expressions. Problems arise when we externalize our reason for action and objectify the result we pursue. We become more exposed to the drifts and seductions of persuaders, because we fail to see the particular propensities we have developed as a result of our phylogenetic history. Technocapitalism makes us believe that our passions are the consequence of shortcomings. We therefore find ourselves trapped between two alternatives: to follow the propaganda that urges us to consume, or the mystical path of the liberation from passions. We should follow a third path instead: not the liberation from passions, but the *liberation of passions* from the misleading idea that they are triggered by external reasons or the lack of something. The liberation of passions requires us to know what the predicative characteristics of the human animal are. We must become aware that passions arise from expressive and copulative needs and not from appropriative needs. Let us clarify this with an example. We humans enjoy collecting. To collect is a constitutive predicate that makes us easy prey to the seduction of marketing. In order to free ourselves from compulsive purchases, we should understand that our desire is fulfilled by performing the action of buying rather than owning the object.

We deny our predicative nature if we assume that our motivation to buy is triggered by the actual value of an object. Thus, desire is twisted and interpreted as deriving from the lack of the object itself. Because humanism denies human animality not in biological terms, but in terms of ranking (Gehlen

1988), humanism also assumes that, unlike animals, we are not guided by inherences as expressive orientations (Leopardi 1997). In other words, for humanism, there are no directions (predicates) guiding our choices; humanism assumes that we are totally free. Thus, people are easily tricked by the smokescreen of the result. If we are not fully aware of our own nature, we get easily and subtly manipulated. We are made to believe that we desire these addictions, hence we become captors of ourselves. Let us take another example. Humans do not like wars, but they may become easily infatuated with this terrible experience (Eibl-Eibesfeldt 1979) by: (i) directing their competitive and affiliative tendencies toward nationalistic behavior; (ii) becoming dangerously addicted to the euphoric effects of adrenaline; and (iii) directing predatory behavior toward fellow humans. Liberating the body means exposing the mechanisms of everyday psychopathology, which transform institutions into organs of hidden biopower appointed to control and direct (Foucault 1977).

The liberation of the body requires full awareness of our own tendencies. It is a Delphic principle that urges us to go past the relativism of ranklessness and totally acknowledge our being-animal, namely our phylogenetic legacy. To be conscious of this means abandoning all forms of determinism and emancipating ourselves from dependencies (Leonhard 2016). Above all, it means being less exposed to deception and circumvention. No freedom is possible without awareness of our nature. Acknowledging that there is something like a human nature does not mean falling into hereditary determinism. Our phylogenetic legacy is merely the building material, not the finished product. It is a *copulative principle* (Kristeller 1943) and a flywheel of experience, not the result of experience. Rediscovering one's corporeity involves rediscovering the uninterrupted thread that connects us to the biosphere, rejoicing in feeling part of it, experiencing conviviality with otherness and understanding that no existence is possible in separation and disjunction.

The first step toward this conversion is to shift from the notion of *Having-a-body* to that of *Being-a-body*. *Having-a-body* refers to the principle of hospitality, whereas *Being-a-body* means that we cannot separate being from body, disconnect subjectivity from corporeity. Living in somatic fullness does not mean indulging in hedonist excesses, but considering one's time and one's presence as a feast, valuing and cherishing our body, offering it as a gift and enjoying this participation and conviviality. Life is a feast, and we humans, through our presence, become patrons of encounters and organizers of feasts: this is what I mean by full somatic presence, it is a presence made of bodies that meet and exchange gifts (Marion 2018). It is not about leading a life that is deontologically wise, but about seeking happiness in wisdom. Regardless of some similarities with some great thinkers of the past, posthumanism is a philosophical revolution. For posthumanism, immanence is not a fallback. The goal of posthumanism is not to amend some of the characteristics of immanence, such as finitude. Its goal is rather to celebrate immanence as participation and, especially, as transience, as well as to avoid humanistic solipsism, which looms over us with the shadow of death.

By looking for the meaning of existence inside an isolated and self-centered individual, humanism fails in its intent and prompts a sense of loneliness, harbinger of despair (Schopenhauer 2010). This tradition has a pessimistic aftertaste, which is perceivable even among authors who are not so explicit about it. The solipsism of the individual withdrawn in his or her *cogito* fosters the feeling of *Thrownness* and becomes a *modus operandi* that reduces the space of con-philosophizing. When reading the philosophical works of the humanist tradition, in fact, we cannot help noticing a basic flaw: self-referentiality, namely the idea that, rather than opening up to a broader discussion, the human, and the philosopher par excellence, ought to remain inside the sacred borders defined by prejudice. Self-complacency prevents him or her from learning anything from his or her encounters, even with the smallest of insects (Bruno 1998). Humanism seems cramped, detached from reality. It somehow resembles echolalia, different philosophers doing nothing else than reporting and commenting upon each other. Posthumanism, on the other hand, aims to revive the lofty function of philosophy, which is to interpret the relationship between the human and the world. It seeks to do so by overcoming barriers that are not only disciplinary, but that post-sophistically refer to the sole human dimension.

Bringing our focus back to the body means revisiting some themes and reversing previously drawn conclusions. If passions are predicative expressions (Canguilhem 1989) triggered by inner motivations rather than by objectively desirable entities, if it is not scarcity but expressive exuberance that stimulates desire and, if the latter is copulative, because it works like a verbal predicate, we must consent that, rather than limiting our freedom, passions are flywheels of freedom and motors of creativity. If our feelings sustain our values (Bodei 2018) and affect our moral actions, reducing the risk of inattention and sloth; in other words, if our emotional bonds are calls to ethical proactivity (Levinas 2000) and our sentience is not limited to suffering; if our emotional bonds are elements that bind us to the world, it clearly makes no sense to claim that our choices are based on rationality alone. When I argue for our need to gain a new awareness of our *Being-a-body*, I mean that *Being-a-body* is not being imprisoned or exposed, but participating and feeling involved. The body is free whenever it can take part in something. Being a free body means becoming aware of the body's motivations, hence becoming emancipated from the ignorance of its reasons. The body can eventually breathe, understand its possible conjugations. The body's multiple relationships with the world and the cultural dimensions it has interpreted, including the technopoietic dimension, make these conjugations possible. The technosphere is no longer a shell that contains a body, but a set of connective elements that extend the banquet.

## FOURTEENTH THESIS

Posthumanism is not anti-humanism. It does not subscribe to the idea of the programmed extinction of the human and is aware of the importance of human action in remedying its errors and reconstructing a new way of life.

Posthumanism is often accused of being an expression, not even thickly veiled, of anti-humanism (Foucault 1970). It is often accused of demeaning the human, even of proposing policies that narrow human prerogatives. This betrays a misunderstanding or just a superficial reading of the terms of the posthumanist proposal. First of all, I believe that whoever undervalues and shows no concern for our serious ecological crisis is the true enemy of humanity, because he/she curtails the future of the forthcoming generations. Anthropocentrism is not a bastion in defense of human development. On the contrary, it hinders the development of our predicates and threatens the survival of our species. It is an obstacle because it does not view human predicates as copulative entities that have developed in correlation to the environment and, therefore, it disconnects the human from its dialogue with otherness. It is a threat because, in its self-justifying folly, it overlooks the impact of human actions on the biosphere. Anthropocentrism is therefore extremely dangerous for our future. Posthumanism wants to remove this obstacle and free the individual from forms of limiting narcissism and egocentrism (Han 2017).

Even though some of our tendencies and propensities are clearly a risk factor, the problem, as I said, is not posed by the human as such. The problem is the cultural and social framework called upon to conjugate our tendencies. Not all human communities have the same impact on the biosphere, nor can we possibly compare Neolithic communities to the industrial and capitalist system of the last few centuries (More 2016). The target of posthumanism, therefore, is not the human but the particular model that has produced major problems, such as imperialism, colonialism, and capitalism, and their consequences and offshoots, such as globalization and the concept of the zone of influence. Posthumanism neither presents itself as anti-humanism nor anticipates or advocates for human extinction (Weisman 2008). Posthumanism offers a new model that: (i) delves into the most blatant expressions of the human drift, such as consumerism or the accumulation of goods and power in the hands of few elites; and (ii) targets the paradigmatic foundations that brought this all about: its goal is to go beyond a specific social and cultural, hence economic and ecological, model.

The concept of human drift lies at the heart of our argument (Rosa 2010). We need to focus on our progressive, linear, and exponential parting from our center of gravity, whereby I mean our inner cohesion and relationship with the world. This drift is like a global psychopathology, a disease or deterioration of the human dimension. The human dimension has been scattered into impulsive drives that are an end in themselves. They have no inner balance or comprehensive integration; they do not hold together or adhere to the world (Debord 2021). Because humanity is rotting, the problem we witness is not due to the human, but to its deterioration. A relationship with the Earth in all its expressions is missing. Our every action is not one of participation but of distancing, devoid of bonding and roots, of the feeling of experience (Heidegger 1927). What is missing is that *emotional relationship* with which we should be familiar since our childhood. People no longer see plants, insects, and birds, but only color splashes; rather than listening to chirping they only hear buzzes, and even the scant presence of trees is something that annoys them. What I mean by "drift" is this parting from nature that is responsible for human self-referentiality.

Hypnotic visual practices—cinema, television, and, nowadays, computers and smartphones—are important experiences that I do not mean to stigmatize. If they take over us, however, they may become hallucinogenic and toxic (Pallavicini 2020). They prevent the individual from coming into contact with the nonhuman world, because all stimuli are mediated by the anthropoplastic process of homogenization (Rosati, 2017). This is particularly evident in our relationship with animals, which is dominated by the Disneyan model of the mask and the falsifying and glossy aesthetics of neoteny. Analog media permitted mediated experiences, experiences that involved concrete and situated acts: touching the vinyl of the record when listening to a song; smelling the acid and seeing the red light in the darkroom when developing a photograph; placing a film on the projector when preparing its screening and feeling the paper between the fingers when reading a book. In the digital dimension, all this translates into an experience that is: (i) immersive or removed from context, dislocated and disjointed from spatial and temporal references; (ii) without experiential specificity, because watching a film, listening to music, and reading a book involve one and the same experiential dimension, that is, digitalization; (iii) one-dimensional from a perceptual and operative viewpoint, because it is predominantly visual: we only need to touch a pad (phone, e-book, or computer). Therefore, the three experiential dimensions previously described disappear. Digital devices are replacement mechanisms that reverse the very meaning of sensation: from a phenomenon that connects us to the world, sensation turns into a strategy to disconnect from the world and withdraw into ourselves. Moreover, in a digital dimension, experience is enhanced by hyperrealistic artifices specifically designed to increase self-referentiality and ease. Execution is simplified, less grounded and profound, demanding less engagement and commitment.

The drift consists of the harmful withdrawal of the human into itself (Rosati 2017). We lose familiarity—hence bonding, attention, and interest—with everything that is not thrown into the meat grinder of anthropoplastic uniformity. The human as such can only indirectly be blamed for a similar distortion. Since we are vulnerable to the factors that influence our growth and are subject to the experiences we make when we are growing up, the loss of these elements must be imputed to our ontogenetic development, to factors that have emerged along our ontogenetic path and have shaped and molded us. In other words, since redundancy and an infinite possibility of conjugations make the human system virtual, we are also very sensitive to external prompts. It is our species' strength and copulative talent that make it fragile. We are subject to the educational directions we receive. If we grow up in nature, we will develop sensitivity toward it; if we grow up in an artificial world, our sensitivity toward nature will become deficient. We used to imitate biodiversity and be interested in natural phenomena, but today we have become victims of the narcissistic traps of the psychedelic mechanisms we ourselves created (Ritzer 2014). We are the fruit of a process that has detached us from the world, and we have begun to orbit around ourselves. The early humanists believed that this was the only way in which they could flee the darkness of the Middle Ages and restore the meaning of Terence's statement: "*Homo sum, humani nihil a me alienum puto*" (I am a human, and nothing human is strange to me) (Terenzio 1963).

Self-centeredness and the desire to transform the nonhuman world into a flat backdrop against which human three-dimensionality could emerge and stand out carried with it the seeds of today's critical issues. As repeated many times, humanism is a project of verticalization. But verticality produces detachment, distance, and loneliness, as evidenced by today's tendency to drift away from the biosphere. Hypnotic mechanisms imprison the human in a mirror maze (Descartes 1970). They create the illusion that any distorted reflection we see is otherness, while being, in fact, nothing but an altered self-image. The human drift consists of this continuous backward-looking perspective. As Rilke says in his eighth elegy, the animal beholds and faces openness, whereas the human can only look back, its gaze directed inwardly rather than outwardly (Rilke 1993). Before arguing that disjunction has been fully accomplished, we only need to wait for: (i) the human to have sunk into a total and permanent anthropoplastic dimension; (ii) the human to have been connected with virtual realities and metaworlds through nerve connections that bypass even perceptual transducers; and (iii) technological supports to interfere even with our affective and cognitive psychological states (Gawdat 2021).

But it would be wrong to make it merely a question of technology. What matters is not so much whether we get to the result by means of the *black mirror* of social media or psychotropic inducers. What matters is that substitution processes make disjunctive psychopathology a habit, an existential project (Spitzer 2012). The subject is detached from the world and its components disassembled, with a resulting effect of permanent schizophrenia (Greenfield 2015). Even though this process is reinforced—and, somehow, revolutionized—by digital technologies, its roots are to be sought in the past. In other words, its programmatic direction predates the advent of digital technologies by a few centuries: it is the elevating and emancipatory idea of the human. What is really anti-humanist is the withdrawal of the human into itself, and certainly not the challenge that strips the human of its most important prerogatives by revealing that the basis of anthropopoiesis is its copulative capacity, its openness toward nonhuman partners and propensity for hybridization. We could argue that a proposition is anti-humanist when it stands in the way of the anthropopoietic mechanism conceived as a relational and hybridization process, tapping into human relationship with the world as a source.

Psychedelic mechanisms do not replace but invert the very meaning of the phenomenological principle. This happens regardless of whether they rely on the alteration of neurobiological processes or work on perceptual illusion and, through neurosimulation, realize the Cartesian pathology of the illusion of the world (Leary 1964). Evolution has provided us with mechanisms that make us relate and interact with the world. Paradoxically, we are using them to obtain the contrary, such as narcissistic isolation and self-withdrawal. Thanks to evolution, we have developed neuromodulators such as dopamine and serotonin. The former boosts interest in and curiosity for the outside world; the latter stimulates industriousness via the mechanism of fulfillment. Today, the same results can be obtained through molecules inhibiting the reuptake of these two neuromodulators. Something similar is true for pleasure that is stimulated by exogenous opioids rather than through physical activity and the release of endorphins. All these mechanisms tend to increase an individual's solipsistic tendencies (Simmel 1971). This is what leads me to argue that, when it comes to the relationship between the individual and the world, the humanist presumption of our superiority over the rest of the living only produces loneliness. But the same is also true of our relationships with our fellow beings: they are more and more like simulations.

Our individual sense of responsibility collapses. We are so desperately entrenched in ourselves, so defensive, always alert and even wary of relationships, that we can no longer sympathize with others. This miseducation, our incapacity and unwillingness to dirty our hands with soil, to watch nature in wonder and awe and to be moved by observing a flower, pose a real threat to humanity. They provoke wicked infantilism (Gottschalk 2018) and dangerous narcissism. They extinguish all forms of solidarity and kill our capacity to take responsibility for what is happening. They transform the human into a huge dissipative black hole. All this is what we should worry about and what should urge us to take action.

## FIFTEENTH THESIS

Posthumanism questions the concept of universality, rejecting the homologative and discriminative hypothesis it implies, instead proposing, as a true ontological revolution, the concept of prospective plurality in being and becoming.

Posthumanist philosophy cannot be considered merely an interpretation of the human metamorphosis in the hypertechnological era of information technology and biosynthesis. The critique of humanism is directed first and foremost toward the concept of "universal" and, consequently, toward all circumscriptions and homologations, even when disguised in forms that claim to empower or change the anthropic dimension. This is where the posthumanist proposal most resembles humanism in its original intent: the aspiration to embrace all domains, but in a pluriversal logic. Posthumanism nurtures a new sensitivity toward the ontological dimension of entities. This new sensitivity also affects artistic expressions and every single inflection of creative experimentation (Wolfe 2022). To shift from a universal view, based necessarily on inclusion or rejection, to a pluriversal one means acknowledging the importance of diversity and consenting that a different outlook on the world offers an opportunity for dialogue and sharing.

It is no coincidence that posthumanism was a concept born in the artistic domain. It reveals two profound changes in feeling: (i) the cyborg as a particular and unique existential condition; (ii) the theriomorph as a hybrid entity that welcomes animal otherness. Because it is mutant, the cyborg can display emerging characteristics and inhabit new ontological spaces. The theriomorph, on the other hand, is no longer a human enhanced in zoomorphism but an inhabited existential dimension. While the Vitruvian manifesto symbolizes the search for a universal and translates the Terentian precept into art, posthumanism follows an alternative, and somewhat opposite route: it reveals a *pluriversal dimension.* Like light filtering through a prism, posthumanism refracts and multiplies the universal beam into all its possible frequencies, thus legitimizing the ontological existence of them all. This yearning to create an ontological ecosystem or inhabitable, in-between worlds protects diversity and interdependent connections (Levinas 1998). It also produces the need to nurture a new type of sensitivity that does not turn diversity into a monstrosity or view phenomenological appearance as a fixed identity deriving from opposition. This new sensitivity nurtures, dignifies and protects all the different identities.

In all its forms, art can contribute to widening the posthumanist perspective. Thanks to cyberpunk currents and some proposals of new age syncretism (Ferrai 2013), this itinerary was implemented in the 1980s. Our response and sensitivity toward the idea of a nonhuman perspective was affected accordingly. Not just the animal perspective, however—which has been neglected by transhumanism—but also the perspective of replicants or artificial intelligences. Contemplating the multiplication of perspectives, namely acknowledging otherness' subjectivity and the fragmentation of human perspectives produces an ontological revolution that is difficult to take in. This is why posthumanist assumptions suffer from the domesticating tendency of humanist philosophers. For them, as in the case of evolutionism, a pluralism of perspectives can be referred back to dualist categories. The anthropomorphization of otherness—including alien entities, robots, and artificial intelligences—as well as the trivialization of natural entities, demonstrate that humanist philosophers are incapable of going beyond a humanist mindset.

Biology has certainly made a fundamental contribution to the fragmentation of perspectives and the metamorphosis of the universal principle. This is proven by: (i) the gene being no longer considered an essential expression of the species, but a transferable, hence stateless, entity; (ii) the hominid family being viewed as no longer developing along a linear chain but expanding like a bush with the co-presence of different geological epochs; (iii) cognitive ethology overcoming the rigid separation between humans and nonhumans and concepts previously applicable only to humans, such as consciousness, intelligence, and culture; (iv) the development of technologies capable of correlative and elaborative behavior that can interact and compete with humans. This change of sensitivity has dissolved the solidity of anthropocentrism that was based on the ontological principle of universality. This has happened during a period of latency or crisis of humanism that extends from the last decades of the twentieth century to the present. Artistic experimentation preceded philosophical reflection by a few decades (Deitch 1992). Indeed, the latter has held on to humanist points of reference and only embraced the outward characteristics of the posthuman.

We are witnessing an attempt to correlate the emancipatory aspirations of posthumanist pluriversalism with the liberation claims of 1968. This is a mistake, however, even though we certainly owe a debt to this movement that broke through the centripetal immobility of universal ordering. Posthumanism, in fact, is not a liberation of the human, but the emergence and affirmation of the pluriversal, namely of something that transcends the human as such, or the human as a universal. Any attempt to bring back or force posthumanism into a humanist framework means falling back into the soteriological and ascensional tendences of humanism. In order to understand posthumanism, we must clarify what the "perspective fragmentation of the universal" actually means. Otherwise, we risk interpreting the posthuman in a hyperhumanist sense, which is, unfortunately, what we observe in the transhumanist logic. For the latter, the universal is simply the technologically empowered human. This is why transhumanism demonstrates a total lack of understanding of otherness' perspective.

Roy Batty's final monologue in *Blade Runner* and Matthew Barney's *Cremaster* disclose a much more advanced posthuman sensitivity than what we have witnessed in philosophical discussions over the last decade. Even though produced in the 1980s and 1990s, they display a very incisive metamorphosis of sensitivity. Notwithstanding their simplifications and naiveties, the avant-garde reveal the importance of acknowledging a plurality of perspectives and existential possibilities (Braidotti 2013) transcending merely quantitative—hence homologated—concepts of performance, such as we find, for example, in enhanced intelligence. The latter is still pervaded by the Aristotelian-hierarchical logic and its ordered and pyramidal structure of otherness. On the contrary, the cyberpunk cyborg is not merely an empowered entity: it is a refugee who is migrating between the cracks of possible worlds, burdened with the load of suffering and improvisation of whoever must invent its own existence. The cyborg is a fragment of a morpho-functional constitution, who, rather than creating new categories and taxonomies, destroys them, and who discovers affinities not by means of similes or affiliations, but by creating shared ecosystems and cohabitations.

The cyborg is not a superhero. It turns out to be vulnerable and open, showing how every strength always has a fragile counterpart. The cyborg does not celebrate solipsism, a typical humanist vice based on vertical leave-taking (Sloterdjik 2013). In a posthuman vision, might lies in fragility and the need to be connected. Similarly, the theriomorphic is not the animal who is locked up in the zoo to be observed or the monster to be exhibited at the circus or the teratomorphic on which to make experiments. The theriomorphic is the meeting point between two gazes that, by reciprocating their perspectives, reach an epiphany. The theriomorphic is not the physiognomic zoomorphic; it is not the mythological or the totemic. It is rather the Dionysian element, which permeates life phenomenology (Henry 2008). It is the expression of sharing and mutation that characterizes the living. It is therefore an expression of encompassment, but also of disruption (Bergson 1911). In other words, the theriomorphic is the expression of a trans-specific channel that neither homologates, nor separates. It denies all barriers and defeats the prejudice of untranslatability, but also of the idea that translation is just a matter of language.

The theriomorphic tells us that for every gaze, there is an observer and an observed (Derrida 2008); that every otherness has a mirror dimension; that there is no resemblance without empathy or interpretation. But there is also another aspect that risks being seriously misunderstood: the role played by technology in the development of the pluriversal. Perspective fragmentation is not the result of the metamorphosis enforced by technology, but of an awareness favored by technoscience because: (i) subjectivity is something other animals have always had; ethology did not bestow it on them, but simply brought it to light; (ii) human history has not progressed linearly from the ape to the modern human. This view, which we might describe as the "Vitruvian revised in the light of Darwinian theory," still informs all the representations of human development. Paleontology instead has introduced the model of the bush; and (iii) technology has always had a hybridizing effect, but information technology has made it concrete. Therefore, I would like to underline once more that the posthuman expands our awareness. It transforms a vertical and ascensional impetus into a horizontal impetus pushing toward ontopoietic conviviality.

Posthumanism leads to a radical shift in all fields of thought, from aesthetics to epistemology. It is an ontological shift far more complex than a fad, an attitude that makes everybody claim today that they are posthumanists. Posthumanism changes paradigmatically our concept of perspective. Yet if we approach the posthuman as we do a Japanese manga or a North American superhero, who is shielded (solitary) in his hypertechnological armor and motivated by the same ascensional and soteriological goals that informed humanism, we will entirely miss the point. A portrait of the posthuman that simply replaces superpowers with technological grafts is merely an anthropomorphic image that trivializes rather than unveils otherness. It is not a disconcerting but a reassuring portrait because it does not introduce anything new and, most of all, because it does not require a paradigmatic reversal, such as the metamorphosis of the universal into the pluriversal. The sci-fi facade simply restages the humanist approach in disguise.

In the posthumanist perspective, technology is more like a detector and a litmus test than a mutagen and lends itself to being seriously misinterpreted. This is because we tend to focus on the equipment instead of the fragmentation of perspectives. Accordingly, we tend to view the posthuman as the evolution of *Homo sapiens* through new technologies. Posthumanist philosophy is not at all easy to understand, for two main reasons: (i) it is counterintuitive; it opposes the anthropocentric perspective, which tends to standardize points of view by ignoring the otherness' perspective. This happens either because otherness is co-opted and anthropomorphized, or because it is reified and denied the possibility of having a perspective; (ii) it opposes the universal framework of humanism along with the ideas of uniqueness and elevation that apply to everything that is human-related.

The posthumanist shift will require a transition period of a few centuries in order for the human relationship with otherness to be reinterpreted. Words fail me to describe and underline the unease I feel whenever I observe how posthumanism is portrayed: technological celebration and anthropocentric persistence simply emphasize human loneliness. In the definition of the anthropocene and in the technological shells called upon to represent the human of the future, caught between artificial intelligences and anthropomorphic robots, the dominant attitude is solitude. The embarrassing loneliness of a human imprisoned in their own self-reflections is what we witness once again—all mirrored by the narcissistic hypnosis of anthropoplasty. On this horizon, animals' absence is consistent with the humanist verticalization and its estrangement from nature. What prevails is the technological kaleidoscope, anthropocentric fascination, an attitude of superiority and arrogance toward a nature that is considered puerile and obsolete.

The absence of animals in these representations is a disarming demonstration of how big the anthropocentric *hybris* is. As is noticeable in the dystopias of some science fiction authors, with the exception of Philip K. Dick and a few others, animals are the great absentees. And whenever animals are present, they are simply trivialized or anthropomorphized. Whenever the posthuman is interpreted in ways that stick to technology and the human, nature is demeaned. This self-complacency does not break away from humanism but, on the contrary, reinforces it. It would be more correct to call this attitude "hyperhumanist." After all, it is much easier to discuss technological ellipses and technomorphized posthumanities than to question prospective and ideological anthropocentrism. By doing so, all changes remain apparent; everything changes so that nothing changes. Discussing a plurality of perspectives and existences actually calls for a propaedeutic reflection upon the concept of animality: this should remain a firm point. What should also remain clear is that it requires us to question the mechanisms of homologation and distancing that have made of animality a segregation archetype.

# *Chapter 5*

# Projects for a Posthuman Era

The posthuman project not only questions the humanist approach, it also makes new proposals for the next generations and for future societies (Wolfe 2009). We discuss the posthuman era not in order to fantasize about post-organic humans and artificial intelligences, or to indulge in science-fiction landscapes, but to define what the emerging needs of tomorrow's humanity are (Shaw 2017). The four theses presented in this chapter introduce new points of reference for civic education and civic engagement and redefine the very concept of citizenship. The posthuman era will have to deal with wide-ranging critical issues such as our relationship with nature, the definition of spaces, technical evolution, and multiculturalism. We need to revise the role of the intellectual as well as our educational principles, which must become more flexible and adhere to everyday life. Our educational model will also need to change, and posthumanism proposes a *new sentimental education.*

## SIXTEENTH THESIS

Posthumanism must be translated into a movement that spreads across all fields of human expression, beyond philosophy alone, resulting in a global project capable of revolutionizing the social structure and the very concept of citizenship.

The posthumanist proposal does not limit itself to a criticism of humanism but takes into account the consequences that this cultural climate has produced. This means assessing the social and economic models as well as the models of citizenship that define the institutions that regulate cohabitation and participation in the ecumene (Pietrzkowski 2018). Humanism should not be considered only from a theoretical viewpoint as an anthropological lens that offers a particular interpretation of the human. What mostly distinguishes humanism is its purposeful nature, its vocation to design opportunities for the sprouting of a new human. In other words, humanism made an epochal impact and laid the foundations of the modern age. It prepared the modern age through human emancipation. For a certain period of time, humanism helped liberate the human from all forms of subjugation from theocentrism, from the tradition, from an ahistorical and absolute interpretation of Aristotelianism (Garin 1958). As a project, humanism presents itself as a reform that defines the preconditions for all future changes: it urges us to observe reality with a critical spirit, for example to read texts without the mediation of other texts and to search for lost works and their original versions.

The various forms of subjugation rejected by humanism included the subjugation to nature. Perceived as extension, nature was desacralized. In line with trends already visible in the late Middle Ages (Montanari and Andreolli 1988), the human felt that it had the right to make free use of nature and, I would say, exploit it unconditionally. For humanism, the human was the measure of all things, an entity that subsumed the whole of reality. It also transformed into a sort of golden ratio, anticipating and justifying anthropoplasty. Humanism became a project. As such, it was not limited to either the artistic or philosophical field, instead, its applicative consequences radiated toward all sociocultural sectors. The posthumanist critique of humanism needs to forensically analyze the latter's premises rather than just criticize its excessive celebration of technology. While it can be easily justified because of technology's kaleidoscopic effect, this response does not help to go beyond the humanist model. A critique of humanism needs to reveal how inconsistent the idea of a liberation from nature is; it must show that it is a major blunder. While other processes of liberation have contributed to human emancipation, in fact, the liberation from nature has imprisoned the human more and more within ideologies and utopia (Servier 1967).

Reversing the view of the human as the meter and subsumption of everything means rejecting anthropoplasty along with its homologating tendencies, which transform difference into contralaterality (Deleuze 1988). Posthumanism is a philosophy that confers ontological consistency and value to differences; it removes them from a dialectic of opposites (Deleuze 2006). In a line of continuity with Nietzschean perspectivism (Nietzsche 2020), posthumanism rejects the anthropological reductionism of reality. Indeed, we must admit that using an anthropocentric focal lens to convert nature into an idea is a simplification: compared to their originals, all representations of nature are diminutions, human metabolites that have been homologated. Posthumanism, therefore, draws inspiration from the emancipatory project of humanism but revolutionizes its premises. It starts from the assumption that human freedom is full participation in nature, that human freedom is an *emancipation of nature* and not from nature (Jonas 1966). Posthumanism becomes a project that can reconnect the human to nature as well as nature to the human. Its aspiration to emancipate nature became possible through human ontological conquests thanks to reflections conducted upon *res cogitans*. These do not relapse into anthropomorphism because they do not apply only to the human but take the metapredicative qualities of *Being* further than the human, they apply to life.

While humanism should be viewed as a cultural climate engaging all fields of knowledge and creating opportunities for the flourishing of a new human model, posthumanism stresses the need to encourage and support the mutual connection between nature and the human as a global project. This obviously does not mean reverting to oleographic representations of a harmonious and static nature, or to a preadamite condition or to the Golden Age. It means a *naturalcultural* (Latour 2004) revitalization of the human dimension. Posthumanism pursues an epochal transformation motivated by the realization that the contemporary dramatic situation badly needs to be remedied. The posthumanist project is based on the concept of *Eutopia*. Eutopia uses organizational points of reference that are present and possible in nature, rather than being based on an anthropocentric dream. The guidelines of this project, which, I repeat, should not be confused with a return to Walden, that is, with neo-Luddite conservatism, reveal and stress the importance of learning from ecology in order to bring about compatible social models. These new models use cultural tools developed by modernity, including technology, to facilitate the transition toward new ways of being in the world.

Posthumanism, therefore, cannot be interpreted as a rejection of the modern age. It is actually a way to go beyond it, to take advantage of our immense philosophical, cultural, and techno-scientific legacy. For example, without Darwinian evolutionism or the Heideggerian concept of *Dasein* (Heidegger 1927), what could our concept of animal subjectivity be, a concept that has been so important for the understanding of human conjugation? Without the tools of technoscience, how could we possibly find a remedy to the ecological disaster, or how could we address the epidemiological crisis, without the achievements of biosynthesis? Posthumanism does not reject modernity. It avails itself of the conquests of modernity to go beyond its models, like an electron that, having accumulated enough energy, jumps to a larger orbit. The convergence of fields of knowledge requires a new type of intellectual, capable of moving across disciplines, not unlike the way the forefathers of humanism did. What we also need to do is to reflect upon how to translate the posthuman project across different disciplines and avoid clinging to the philosophical debate only, as this would contradict the principles of posthumanism itself.

Posthumanism aims to inaugurate an epochal transformation and promote a new model of existence, starting from the domains of education and citizenship (Snaza 2016). This model is based upon coexistence and convivial participation instead of individual self-accomplishment and human supremacy over nature. As previously stated, it is a revolution in values that creates a shift at several levels: in our way of living life as a *convivium*, in the need to define new goals and purposes and establish new and innovative models of social coexistence. In addition, the posthuman project needs to create opportunities that will allow the new model of human to prosper. This will also help us relinquish attitudes that produced loneliness and hostility toward otherness because we felt the need to take revenge for our inner suffering (Demichelis 2018). Key to this change is to abandon competitive individualism and existential solipsistic withdrawal (Bauman 1998) by finding new meanings to our presence in the world that are characterized by openness and participation. Essentially this means: horizontality of the subject. The subject recovers a sense of existence by stretching out its arms to the world, like a new Vitruvian, who embraces the world rather than being a measure of the world. For this reason, posthumanism cannot hold together without projectuality.

If we analyze the humanistic precept of verticalization, we realize that it not only applies to the theoretical domain, but also to practical everyday life. This is where the artificial, the tendency to promote anthropoplasty and give preference to nonmaterial work prevail. In our everyday lives, we have detached ourselves from the earth and become estranged from nature (Nietzsche 1917), we have lost familiarity with the dimensions of the living and become profoundly ignorant of all that pertains to the animal and vegetable otherness. This split is the root of negligence, a form of affective forgetfulness that is observable in how we deal with the ongoing ecological destruction. Negligence derives from a sentimental miseducation that alters value structures and affects people's priorities. Separation not only nurtures carelessness and directs preferences toward human products. It can also be observed in oleographic, and therefore static and disjunct, visions of nature, which feature in conservatism (Macaulay Trevelyan 1942). The same is true for some expressions of neo-Luddism or of nostalgia for a wild and mythological past. Posthumanism believes that the human lives in a metamorphic and dynamic dimension, and that it partakes of this dimension through *rites of contamination*. Contamination means that we are ready and willing to dirty our hands in soil and expand our horizons toward natural phenomena and hybridization.

In order to achieve this, however, we need to build a new life model. It is a model that frees the subject from all the conditionings of the humanist civilization, such as unbridled competition and the need to be surrounded by systems of meaning that pivot around the individual encouraging loneliness

(Simmel 2020). In the adulterated economic-productive model we live in, the Self has become alienated by labor and consumerism. Work has lost its existential meaning and commodified the individual, who is pigeonholed inside the system. The idea of a circular economy does nothing but increase inequalities by peddling the illusion of self-realization (Han, 2018). The individual is forced to play the role of a spectator: private spaces are denied and the individual is deceived into believing that he/she is the protagonist on global stages that actually do not belong to him or her. He/she is forced to acquire evidence of credibility, such as publications, visibility, followers, money, or prestige (Del Rey 2013). The logic of the balance of supply and demand feeds the false assumption that we are free to make our own choices and purchases. It has become clear by now that this is not in line with an individual's well-being. Besides leading to human commodification, it is a model based on positive feedback mechanisms, which favor inequalities and social antagonism.

It is necessary to build an economy for the people, not the other way around. This is in order to create opportunities that guarantee everyone an effectively solid and fulfilling growth. Being forcefully placed in an unsatisfying system, subject to continuous surrogative processes and doping mechanisms, causes the subject's infantilization. Precisely because of the perverted structure of which he/she is part, he/she floats in a limbo that makes him/her prey to anxieties and compulsions (Lembke 2021). The humanist project could still be maintained during its first few centuries, when societies had not yet reached today's level of complexity and when technology had not yet completely split the human from the terrestrial. This process has acquired a pervasive trend today, with blatant perverse, toxic, and contradictory aspects. The posthuman project has therefore become necessary not just to face the ecological crisis, the loss of meaning, and the challenges of technological development, but also to liberate the human from a self-built prison and self-destroying mechanisms.

## SEVENTEENTH THESIS

The educational aspect is one of the crucial points of the posthuman transformation, because the posthuman dimension must be supported by a redefinition of institutions, the most important being the school system.

Education is one of the core aspects of the posthuman project (Ferrante and Orsenigo 2017). If we continue to base pedagogy on the assumption that we can extract the human from the so-called animality of the child, as if education were an act of alchemy or purification, we should not be surprised if young people behave in a certain way. Such a pedagogical approach is influenced by humanist dualism too: the human is placed outside nature and the animal condition is viewed as contralateral to the human (Cimatti 2013). Education is either a corrective process or appeals to the essentialist principle for which inherences can develop coherently and autonomously only when they find themselves in the right context. Human nature has characteristics of its own (inherences). They are not extracted through a process of purification, though, but should be viewed as masters of experiences that open the human up to ontopoesis. In the humanist perspective, the corrective process is supposed to remove that animal veneer that hinders the complete expression of the human essence and the possibility of it being true to itself. Education is a process that counters animality and imposes superstructures peddled as expressions of authenticity. Education becomes a process of liberation from animal residues because they obscure predicates that need to be brought out (*ex-ducere*).

In the posthuman approach, education is a process of hybridization. It comes about through events of gradual contamination with the world that are facilitated by human copulative propensities. Human inherences do not develop in autarky, and, most of all, they are not supposed to be purified from alleged animal residues. On the contrary, they need external contributions (Barad 2007). The humanist model of ontopoeitic development forces us inside an oppositional dichotomy. It views growth and development: (i) as the independent sprouting of an essence that only needs to be cleansed from impurities; and (ii) as external conditioning that confers shape to a *tabula rasa*, a blank structure devoid of inherences. The posthuman perspective instead relaunches the *dialogical meaning of education* (Pinto Minerva and Gallelli 2004). By providing the individual with precise points of reference for its encounter with the world, the innate pushes the subject toward experiences. But the context of development is never fully indefinite and uncertain. We can refer to a niche inheritance because our developmental niche consists of our affective relationship with our parents, our social structure and cultural dimension. The niche is not a static but a dynamic entity in continuous becoming. Hence, posthumanism must come to terms with becoming.

In the posthumanist proposal, education becomes an initiatory journey into hybridization. Rather than being hampered, this journey is facilitated by our inherent characteristics. This is because human nature: (i) is predisposed to receiving external contributions and (ii) can be implicitly conjugated in different forms that are dependent on the relationships it establishes with the world. From this perspective, culture becomes the individual's ontogenetic niche, the initial whole that makes the biographical subject the expression of previous dialogues. All the external elements that enter this dialogue are introjected, not as simple footprints or traces, but as fruits of dialogue. In other words, while a footprint is something passive, a dialogue implies a mutual exchange that produces new contents and lets them emerge. We ask questions and the world answers, but the answers we get reply to the specific questions we ask. The relationship with the world creates subjectivities. Accordingly, there is no environmental determinism in ontogenesis, because every single answer of the world replies to an experiential question of the subject. I would therefore like to underline the importance of not considering hybridization a trace left by the outside (Bateson 1976), a superstructure enwrapping the individual, or the simple introjection of unchangeable elements. Hybridization is a singular and unique event that eschews all forms of determinism because what gets introjected via a conjugation is always the fruit of dialogue.

What stands out in humanist pedagogy is its tendency to neglect phylogenetic legacy in the individual's copulative orientation, namely, in the experiential questions the subject poses to the world (Dehaene 2020). The learning process is assumed to be rather passive and to rely on external incentives. We forget that what elicits experiential engagement is the intrinsic copulative condition. Let us consider, for example, the key role played by emotions and motivations. Because of the separation between the humanities and the natural sciences in the humanist paradigm, even the physiology of the child, which is so central to evolutionary processes, is almost ignored. It is thus all the more important to reconnect what humanist dualism has separated (Derrida 1982), starting from the dichotomy between science and philosophy. The posthuman epoch will be based on a new pedagogy that: (i) may reconcile human tendencies with the development of anthropopoietic predicates; and (ii) envisages an education in line with the neurophysiological factors that influence learning processes.

A biocentric pedagogy (Naess 1991) informed about, and based on, nature has two main objectives: (i) encouraging inclinations, no longer conceived of as deterministic factors to be respected in a Rousseauian logic but as copulative flywheels of anthropopoiesis; and (ii) increasing the conviviality with otherness in a logic of participation and existential horizontality that may promote the posthumanist dimension of *Being-with*. In order to develop a project that may overcome humanist limits and constraints, we need to start from radically different premises. Posthumanist pedagogy aims to favor the singularity and uniqueness of an individual's experience, hence its irreducible character, and underline its distinctiveness and peculiarity. But this must be done in a logic of exchange with, and openness toward, otherness that does not limit itself to merely demanding respect for the *convivium*, but celebrates it too.

There is a highly problematic aspect in the individual's narcissistic self-withdrawal. It is evidently the consequence of the performative education that pervades the humanistic vision and its acrobatic imperative of verticalization (Sloterdijk 2013). As long as what prevails is the rule of individual affirmation over others, or, should I say, of individual self-affirmation at the expense of others, humans and nonhumans alike, we should not be surprised if this model extends from the subject's microcosm to entire nations and societies. Humanism requires the individual to strip itself of its biological prerogatives in order to reach an extranatural dimension. Humanism expects the individual to transform life into a work of art for its own sake, a work of art that is exhibited to the spectators' cold gaze but can neither embrace nor create a dialogue with the world (Levinas 2000). It is a work of art without solidarity and sympathy, a mere exposure of the self—the phenomenon is also typical of some social media, such as Instagram. This is how even works that are decreed by public consensus to be successful suffer from incurable loneliness: the greater the crowd that gathers around them, the bigger their loneliness. The reasons for this paradox are lack of dialogue, excessive self-withdrawal, and mere exhibition—never truly a gift.

The unbridled individualism of the contemporary period cannot be dismissed as a drift or an aberration. It comes from a specific education to which the child is exposed from a very early age. It is a sentimental miseducation that leads the child to withdraw progressively into itself, to take arms against the outside world and to structure in conflictual terms even its most intimate relationships. It is captious to argue that these traits belong to human nature (Hobbes 1881), because every tendency in nature is always counterbalanced by a contralateral tendency, and every expression derives from how the propensity has been conjugated. While this model rests on phylogenetic inclinations, it is clearly reinforced and conjugated through the points of reference previously described: verticalization and solipsism. In encouraging disjunctive verticalization, humanist education favors not only an ideal direction, but also a particular type of emotional habit. The latter is characterized by a disconnection from the Earth and an estrangement from life in its various expressions, including one's own body. The naturophobia of humanism (Benasayag 2015) is first of all a disjunction of feelings—clearly a miseducation according to posthumanism—a loss of familiarity with life phenomenology (Verducci 2012). Even corporeity is experienced with great difficulty, as if one's own physiology had to be cleansed too.

There is clearly a line of continuity between the transhumanist approach and the naturophobic approach of humanism. Technology provides the means to imagine a shift to an extrabiological dimension (Simondon 2014). From this perspective, the representation of the postorganic human (Savulescu and Bostrom 2009) easily helps us getting around the paradigmatic crisis of humanism. By relying entirely on the mundane dimension of verticality supported by a Promethean apparatus such as technology, humanism can avoid the human's shared and hybridizing dimension. But the excesses of technology, within which the human is hermetically sealed, are not only the results of the technological appeal. They express a feeling deeply rooted in humanism, namely the desire to be protected from nonhuman contamination, the desire to live within an anthropomorphized projection of the world in complete separation from the legacy shared with nature.

Human forms drawn within an epidermis of sheet metal and equipped with entrance doors for telematic systems seem puerile to me. Paedomorphisms that remain aesthetically appealing in a world without parental feelings, or forms of intelligence that can only rely on their elaborative potential because they are devoid of affectivity, betray a total ignorance of natural laws. How can similar representations be said to differ from humanism? I therefore believe that the *brainframe* (De Kerkhove 2019) that does not enable us to transcend these projections originates from the education and models of citizenship people experience in their daily lives. Rather than an ideological choice, it is the consequence of an applicative diretrix that has gradually established itself in Western culture and has later found fertile ground in many other cultures.

Posthumanism believes in the necessity of a *new sentimental education* that may foster the bonding between the individual and the universe of animal and vegetable otherness. Only through a total immersion in the nonhuman universe shall we become capable of affectively bonding with the world (Han, 2022), rediscovering the meaning of our existence and understanding the ontological plurality that awaits us. There is no opposition between technopoietic development and deep immersion in nature. They are actually not separate, but their division only derives from the disjunctive dualism that underpins the verticality of humanism. Culture, and *téchne* along with it, are but one of the manifold expressions or levels of nature. Immersion in nature and horizontal opening toward biological otherness produce an enrichment that has always been the basis of anthropopoiesis and, accordingly, technopoiesis. To keep on reasoning by oppositions means prolonging the humanistic path of the antinomy between the Epimethean and the Promethean (Scheler 2009) and holding on to a biased logic of a prioris that views the two as ontologically separated. I not only suggest that we should promote a wider and deeper knowledge of nature as well as more substantial practical experiences in it, we should also pursue a sentimental education that engages us in an experience of life as a whole, including affectivity and aesthetics. This type of education produces both values and knowledge. It makes the individual bond with the world and share his/her feelings with it.

## EIGHTEENTH THESIS

One of the main proposals of posthumanism is the need for a new kind of intellectual, capable not only of reconciling science and ethics through dialogue and mutual questioning, but above all of hybridizing them and translating the new emerging knowledge into new customs.

If we find it essential to speak of a posthumanist project that may inaugurate a new era, we must also reinterpret the figure of the intellectual. Today, this figure relates to two arbitrarily divided sectors barely in touch with each other, in which everyone is only preoccupied with garrisoning his/her own castle of knowledge. Development in the various disciplines is impaired by the tendency to be self-referential, an attitude that makes the intellectual a cleric devoted to his/her disciplinary priesthood and disinterested in moments of transdisciplinarity. Unaware of what is happening out his/her own garden, he/she performs the function of a glossarist in charge of just prolonging the tradition. The dichotomic model *human sciences vs natural sciences* created barriers that caused the breath of fresh air of the early humanists to wane completely (Snow 1959). The intellectual has become a guardian of the tradition whose contents cannot be refreshed because the one who should do it is sclerotized by conservative anxiety. This intellectual's function is mainly ornamental and makes little to no impact.

The crisis of the intellectual is not due to the fact that this figure has become meaningless, but that it is disconnected from society. The intellectual's role has been reduced to that of a columnist who must comply with the establishment's directives. The expected function of a cleric is to transform political choices into nonnegotiable requests that may endorse the implementation of an increasingly illiberal society (Popper 1962). Confined in his/her own domain of knowledge, the intellectual is perfectly manipulable. His/her task is to transform political choices into directives against which no appeal is possible. Since they require specific competences, they cannot be subjected to the democratic debate. The virologist, the economist, and the geopolitical scholar, to mention but a few, are asked to endorse top-down political choices by translating them into imperatives and prescriptions (Milgram 2003). This operation denies the principle of democratic participation. The transformation of the intellectual into a priest officiating the cult of a particular discipline has favored the evolution of a capitalist society where information is both apodictic and propagandistic (Packard 1957). The jumble of contradictory news facilitated by social media has further fortified the mainstream and caused the figure of the critical intellectual to fade.

Undoubtedly, this development has also been the consequence of the economic and structural acceleration of societies. It is equally indisputable, however, that the fragmentation of knowledge has favored or, at least, been unable to stem this transformation. The progress of knowledge is often said to produce inevitable specializations, with the resulting clerical confinement of scholars. This is, however, a pretentious statement. Nobody can be an expert in everything. But we can legitimately expect a scholar to have a formative background and critical spirit that makes him/her open to dialogue. He/she should be able to take part in a multidisciplinary roundtable (Ferrando 2016) with prior knowledge of new discoveries. Is it today acceptable for an anthropologist to completely ignore human morphophysiology, for a philosopher to discuss animality without knowing anything about ethology and for a teacher to be totally unprepared for the neurophysiology of the developmental age? It is not a matter of knowing everything but of being able to switch between different disciplines when discussing subjects related to them. I do not think it is appropriate for a scientist or a doctor to have no knowledge of epistemology, for a veterinarian to ignore the history of biological thought and for an architect to be totally ignorant of botany.

Only by knowing how to shift between two fields of knowledge while discussing a related topic will the intellectual be able to make a real impact and promote new visions. Posthumanism denies the existence of a dualism (Bruni 2020) between an entity that is extension, nature, which is describable in the status quo and a contralateral entity, the human, which is dynamic and autopoietic, continuously transforming and barely following biological laws. The development of knowledge and its applications indisputably produce specialization. What is disputable, however, is the traditional formative division between natural and human sciences: any topic requires the intellectual to cross this aleatory boundary (Thompson Klein 1990). We do not just need to build a bridge between the two domains, but we must also deny the existence of a gap between them altogether. We must reinstate the *naturecultural continuum* in all contexts of discussion. We must acknowledge that in all phenomena, there is the convergence of the nomothetic and the historical, the double dimension of the general and the contingent. In other words, we must acknowledge that we cannot avoid taking into account both the laws that regulate physical phenomena and the historical contingency of phenomena themselves (Gould 1990).

As observed, this is true in education, which requires knowledge in neurophysiology, human ethology, the psychology of learning and sociology—a background that does not separate the natural from the human sciences. It is at the operational level, namely when it comes to translating knowledge into practice, that the contemporary intellectual fails: trapped in a profusion of erudition disconnected from real problems, his/her propositional efforts, declarations and statements sound like showing off his/her knowledge rather than offering an innovative project (Bauman 2007). The intellectual is incapable of getting into the overall debate; his/her involvement is always only lateral, like an accessory or an ornament. However interesting and fascinating their contributions, intellectuals cannot give answers or formulate the right questions. They always sound somehow off topic. They have moved more and more to the margins, squeezed between a predominance of either the academic or the journalist, two figures who are not always in a position to use their knowledge critically and apply it to society. I am not saying that academics and journalists do not have an intellectual function; I am saying that their main task is different. In the panorama I have described, intellectuals are merely specialists, ill-suited to exposing contradictions, also due to their high specialization, which makes them shortsighted and easy prey to power.

Since we are facing an epochal transition, we need to stress that the absence of the intellectual figure represents a risk for democratic societies: it imposes autocracies supported by the clerics' authority (Habermas 2008). During the course of this metamorphosis, every single political choice is like a medical prescription: it is not open to discussion. The intellectual is one of the antidotes or counterweights needed to prevent homologation and the danger of a single thought when the media's propagandistic incidence and the specialists' prescriptive normalization are high. This is the reason why the intellectual must be concrete and transversal, capable of entering everyday life not only in order to problematize directives, but also to dictate an urgent agenda and define criteria for a broader vision (Benda 1955). By virtue of his/her ability to cross boundaries and discover what hides behind particular choices or directions, the intellectual will be able to anticipate future scenarios and avoid the shortsightedness of both the specialist and contingency.

In order to obtain these results, however, we need to train scholars who know exactly how to reconnect the different fields of knowledge with one another (Morin 1999). If the humanist project separated humanities from natural sciences, posthumanism proposes an ample intervention to reconnect the two domains. They should not simply communicate with each other and stay separate but achieve a synthesis. Each of the two domains will need to avail itself of its prerogatives and contribute to the development of the other. Thereby, the synthesis will not be a simple aggregate but work like an organism, it will look like the spontaneous anastomosis of a plurality of coherent disciplines conflating into an organic system. Humanities in some respects and natural sciences in others have developed highly valuable structures and methodologies. Once again, I would reiterate that posthumanism is not antithetic to humanism, it overcomes it—it is obviously not a continuation. Posthumanism avails itself of the achievements of humanism, without which it could not have developed. I obviously do not refer to the etymological meaning of the word posthumanism, but to the fact that neither its contents nor anthropological evolution could have developed without humanism.

Transversality requires a process of reconnection that is anything but obvious. It relies on the transformation of one discipline's prerogatives into flywheels for the development of the other. That which characterizes the human sciences must contribute to the development of the natural sciences and vice versa, so that their encounter will not produce a simple summation of knowledge but a global development of it. In addition, I believe that the differences that have been accumulating in these two areas over the course of humanism can: (i) favor new emergencies (Johnson 2001), that is, new contents; and (ii) contribute to the two disciplines' mutual evolution and development. I believe the latter to be an extremely fascinating aspect, which might lead to a change we can only partially envisage. When I argue that the posthuman era is one of transition, I also refer to its character of openness and unpredictability. This was also true of the humanistic revolution, whose developments were surely unclear to fifteenth-century scholars. It is up to the posthumanist intellectual, therefore, to undertake this journey across the open sea, toward unknown continents that require an openness and a fluidity that go beyond the tradition's static approach.

Posthumanism will have to set up an ample program of debates in order to support and shape the new intellectual. He/she will need the ability to move transversely across disciplines rather than remaining imprisoned in the microcosm of his/her own knowledge where: (i) philosophers who deal with animality have no idea of evolutionary biology; (ii) didactic programs can be prepared while ignoring the neurophysiology of the developmental age; and (iii) urban planning can be made without the participation of agronomists, doctors and veterinarians. Refusing to be relegated to a cloister, the posthumanist intellectual will learn to be transversal and move across disciplines, an ability that should not be confused with multidisciplinarity. Transversality will result from hybridizing the methods of the two traditions. Only then will the university cease to be an ivory tower and become an *agorà*, while schools, supplied with naturalistic laboratories, will offer opportunities for making practical experiences and technical exercises. The shaping of the posthumanist intellectual will therefore follow a pedagogical model that is completely different from the current one, without losing any of the achievements of the two domains.

## NINETEENTH THESIS

Posthumanism changes the understanding of space, ordinarily divided into anthropic or ecumenical space and naturalistic or wild place, avoiding this separation and seeing it as an expression of humanistic dualism.

The demographic explosion of the past century is contributing to our serious ecological crisis (Ehrlich 1968). We are witnessing an increasing agronomic pressure on the land, which leads to progressive deforestation. Not only is this jeopardizing the existence of residual areas in the most densely anthropized geographical territories, such as Europe and Asia, but it is also threatening the lungs of the planet, namely the large rainforests, which are also the largest source of biodiversity (Barrau 2020). The agricultural conversion of the territory is not only a consequence of the demographic bomb, it is also the result of an energy-consuming eating style that spread first in the West, and then, in the course of the twentieth century, progressively extended to Eastern countries. By "energy-consuming eating style," we refer to the conversion of plant protein into animal protein. This process burns 90 percent of the resources, causing an enormous increase in agronomic pressure, because most of the land is used for forage production. In addition, the development of extensive farming in America has significantly contributed to the destruction of forests that have been transformed into pastures.

A diet rich in foods of animal origin, especially meat (Adams 2015), is not only the basis of many pathologies, but also the major cause of deforestation and environmental pollution. Today, this problem is well known. Unfortunately, it is not easy to modify an eating style that established itself in the middle of the last century, probably as a response to the poverty caused by the two world wars. Similarly, as an expression of acquired wealth, this eating style started to be coveted by emerging countries too. In addition, it is endorsed by scientific prescriptions, in terms of noble proteins, and by the economic value of livestock farming. Therefore, we cannot envisage a solution to the ecological crisis if, alongside the energy question, we do not address our eating style too (Rifkin 1993). We need to consider that demographic development and globalization have led to the crisis of the Neolithic model based on agriculture and livestock. Therefore, the transition requires a change in this particular nutrition and production model. I believe that livestock farming will be the first sector to collapse in this transition period with a sharp reduction in the consumption of foods of animal origin, which are the major cause of deforestation.

Anthropization, namely the territory's anthropogenic conversion, has a wide-ranging impact. Besides the agronomic pressure due to the increasing invasion of natural spaces following the growing consumption of foods of animal origin, agricultural practices are reducing the fertility of the soil (Magdoff and Tokar 2010) through subsidence, fertigation, deep plowing, agronomic overuse, and the exploitation of fragile soils, which risk desertification. The extensive use of herbicides, pesticides, and chemical fertilizers has also contributed to a drastic decline in entomofauna, with a reduction in pollination processes and a subsequent erosion of productivity. Moreover, through extreme meteorological phenomena and desertification, in some geographical areas, global warming has struck the final blow to a productivity that was already collapsing (Diamond 2011). In essence, while humanity's rising numbers and demands increase productive needs, the planet's areas convertible to agriculture are being reduced year after year. This means that, in the long term, abandoning livestock farming practices will not be enough: we will need to accelerate food production processes that are different from those of the Neolithic tradition.

In view of the epidemiological countermeasures that usually accompany processes of ecological overpopulation, it is difficult to anticipate our demographic trends (Randers 2012). We could even witness a pandemic that, in the course of a few years, cuts down most of the human population. Yet, if our techno-scientific capacity succeeds in tackling the epidemic outbreaks that will inevitably occur during this century (Quammen 2012), we will still need to face the challenges posed by population growth: not an easy task, unless we want to turn the whole planet into farmland for food production. Posthumanism, therefore, must find workable solutions to face the difficulties of this transitional period (Stern 2009). These solutions cannot be based only on renunciations, however, because, even if applied correctly by all of the world's population, they would not suffice to solve the crisis. Solutions must also be based on biotechnological developments, which may help us abandon the society model that followed the Neolithic revolution. As a period of transition, posthumanism will also have to equip itself in this sense, namely be able to formulate a *post-Neolithic model of food production*. This transition will certainly occur during this century and lead to a new food revolution.

The destruction of natural spaces is not only caused by the increasing pollution due to plastic and other poorly biodegradable residues. It is also due to other processes, such as concreting and productive conversion, the exploitation of natural spaces for economic production such as tourism. The economic weight of the construction industry in a nation's GDP is significant. This is the reason why it obtains political support causing the further erosion of the few residual and not-yet-cemented spaces. All green corners are progressively invaded by the metastasis of concrete. This phenomenon has also an impact on people's affective and aesthetic sensitivity because they feel more at ease in a home in the middle of a jungle of asphalt than in a forest. The real totem of humanist culture is the ideology of the machine (Mumford 1967), which accompanies the whole parable of the modern age. As a human product, it can externalize human dignity by transforming our species into a demiurge, a new creator. Space is no longer the somatic measure of the human, but of its machinic expression. Cement and asphalt help us convert space into a place consistent with the chimeric and abiological dimension we fantasize.

Our idea of space is affected by the exponential factor of the speed with which we move in it: all places must become a means of transit that can be crossed at high speed (Rosa 2010). In other words, their resistance, their specificity of contrast, must be compressed and reset. For instance, if a territory is flattened in order to build a motorway, the area's resistance and characteristics are drastically reduced. There has been talk about a progressive development of non-places (Augè 2008) in lieu of contexts rooted in the territory. Like impermeable vases, their sole purpose is to transfer someone from one point to another. However, this is only the last phase of the metamorphosing process directed toward landscape destruction that makes space and time overlap. We should not be surprised, therefore, if the ecological crisis is also one of space, with all the consequences it entails, also under the existential perspective of *Being-there*. The ideology of the machine leads us to expect a synchronicity of processes in which verticalization also means being able to exit a temporal perspective (Virilio 1986). This type of perception characterizes the whole of humanism with its obsession with time, as if it were an enemy to defeat. This is also due to the fact that time is resilient to human conversion and human control. When transhumanism displays a posthuman figure floating in an infinite space devoid of references, or immersed in a neurosimulation tank, but also when it conceives of mind-downloading, it does nothing but extract the human from space and time. By doing so, it merely confers a worldly dimension to the spiritual elevation toward immortality and infinity that was typical of humanism.

The posthumanist proposal, on the contrary, seeks to transform this perception into a vision of inclusion, participation and horizontality where the subject aspires to extend its *convivium* rather than reach the *cogito's* solipsistic perfection. The meaning of salvation is flipped: it is no longer an individual aspiration, but a collective process. In its radical enlargement in space and time, the subject builds alliances that confer meaning to its existence and save it from the terror of loneliness and transience. For posthumanism, neither the individual nor the human can save themselves without each other, because salvation is not a lifeboat, but a long rope onto which everybody holds. We must acknowledge that our destiny is shared. This explains why the reflection upon our life space is so important in the posthumanist proposal. Life space is not a *lebensraum*, the archetype of all anthropocentric statements, but a *place of convivium*. Starting from a different planning of the ecumene, our life space extends to all the areas that are directly or indirectly touched by the human presence. We need to start, once again, by overcoming the humanist dichotomy.

The park will not need to be in a separate space, because the entire ecumene will convert into a park. Urbanism and architecture will be the great interpreters of the transition that will give back the park its voice. Indeed, the images of posthumanism do not coincide with those of transhumanism, such as an entity fluctuating in void and darkness, a brain in a tank (Goldberg 2016) or a mind that creates its own oneiric adventures. Posthumanism coincides with a world that enables us to find grounding. *Being-there* is a process of temporal crossing that allows the here and now to emerge precisely because, unlike what the previous images suggest, it is not something outside time or isolated in time. Posthumanism is participation: it means sustaining our presence by relying on the multiple reality levels implied in all encounters. We can never find sense inside or outside ourselves, only in the spatiotemporal convergence of our *Being-in-relationship*. The posthuman transition refreshes the models of life, hence of space, inhabited by the human. These models include a form of planning, which, albeit in a transitional phase, already shows orientations that are significantly moving away from the twentieth century.

This means changing the urban system by, for example, permanently removing cars from cities and transforming streets into spaces where it is possible to move and circulate by other means of transport. Parks will have to become green areas that make alternative circulation possible and every point of the city reachable by walking in the green. Public playgrounds will have to be reestablished; children will need to get more involved in noninstitutionalized activities and will have more free time for direct experiences in nature. The city will gradually fade into the countryside and the interstitial areas between towns. This will be possible through cycle paths immersed in green areas and areas of natural relevance. Crops will have to be transformed in line with greater eco-sustainability and alongside an increase in rural practices that encourage biodiversity. Ultimately, posthuman civilization will have to build a new dimension of space and a new model of existence. The latter will need to: (i) take into account the laws that are in force in ecology; (ii) avail itself of technology in order not to thwart nature but favor the transition toward it, and (iii) optimize conversion resources.

# *Chapter 6*

# The Ontological Question

Posthumanist philosophy questions the traditional model of identity. In a humanistic framework, identity is preestablished and unchangeable, causing segregations and a background intolerance toward diversity (Valentini 2016). For posthumanism, what counts is not merely the politically correct facade. In all matters related to identity, it is important to confer value to fluid and transindividual identities (Simondon 2020) as well as to multiple personal positions. The traditional approach, born out of the assumption that identity is static and unchangeable, defines set positions and roles, dialectics of relationships and hierarchies, and, in general, pigeonholes the individual into categories (Deleuze 1994). The identity revolution of posthumanism acknowledges the value of mutant identities built through bodily experiences and desires. Unlike transhumanism, what this proposal puts center stage is the somatic dimension, a copulative condition always in progress.

## TWENTIETH THESIS

Posthumanism is also a philosophy of the body, a reflection on the meaning of *Being-a-body* and on the character of the cognitive, diffuse and nonhierarchical extension of corporeality, which radically transforms the principle of somatization.

Postumanism is a philosophy of corporeity (Shusterman 2012) in all its implicit and conjugated expressions. The relationship with technology is, therefore, also somatized; the body is modified by the hybridizing referent. To somatize means to include. But it does not simply mean to extend oneself onto the technological device. It also means to reorganize the entire systemic of the organism on the basis of the interface transformations produced by the device. The use of the computer, for example, produces a somatization because it modifies the structure of the neural connectome (LeDoux 2002), the endocrine system, the posture, and the performance of the immune system. Somatization is an evolution of the system. Envisaging an external performativity, the body appropriates and transforms its own performance through its synergic interaction with the device. This means that the body prepares itself to be complemented, to interact synergically with the instrument. Adherence develops through the body. The concept of ergonomics flips because the frenetic development of new technologies causes the body to adapt to the instrument more than the other way around. From an ontogenetic viewpoint, the device produces heteronomy (Marchesini 2017), that is, somatic organizations that expect an external contribution; they are so configured that a higher level can be reached only through the device.

Somatization is a distinctive quality of the body: *cognitive extension*. This means that the body does not have a nucleus or an organizing center but produces multiple and recursive influences: for instance, the nervous system is influenced by the endocrine system and vice versa. Extension refers to a disseminated presence rather than one that radiates from a pulsing heart. It is impossible to find a univocal directionality that puts an organ, district, apparatus, or tissue in control of others (Maturana and Varela 1985). This is proven by the visceral response of autonomous nervous systems: the descending dynamics of the brain-gut axis are a given, but the influence of the heart and stomach on the vagal tone and sympathetic system is equally well known (Porges 2011). Cognitive extension shows the organism to be like an ecosystem, consisting of reciprocal interactions as well as homeodynamic and autocatalytic phenomena. It is a complex system without a center, hierarchies, or a univocal and linear directionality, where each single momentary expression is the fruit of the systemic conditions at a particular moment. Extensive cognition destroys the essentialist image of a shell body that enshrines the spirit. This is replaced by a scattered intelligence (Damasio 2003) that occurs in the subject as an emergence, namely, as a superintendence, without having a precise location.

This image, which causes the two Cartesian terms to collide because it makes the *cogito* a property of extension, dismantles the vision of the *homunculus* as an entity that controls the body by being located inside it or connected with the soul. The body resembles an ecosystem: it lets predicates emerge which make it recognizable and confer identity to it through a complex interplay of interactions. Identity is gained by the historicity of the interactive flow; identity is the history of its interactions and what makes it recognizable in terms of family resemblances (Wittgenstein 1953). The body can thus be viewed as a collection of events held together by a field attractor: its identity is the fruit of historical continuity. The way in which we interpret corporeity changes radically as does the existential dimension and the process of somatization. Somatization becomes a modification of the body's extension and, consequently, of its cognitive emergence. The cognitively extended body is no longer limited to the content under the skin; it is the field of systemic interactions contained in the body's history, exactly like a river that, in full flow, changes its course and transforms itself, swallowing parts of the panorama that it did not previously include.

The ecological reading of the body is counterintuitive. Intuitively, we would be led to believe in a control room, because the emergent effects confer a solid identity to the system, the same that we normally ascribe to the head of an individual. In fact, this is only illusory. It is the consequence of a projection, the result of the same attitude whereby we confer an essentialist identity to a river, mother nature, a forest, natural selection, the planet or, in general, hyperobjects (Morton 2013) (e.g., global warming). By rejecting all forms of dualism, the posthumanist vision helps us overcome the projective or anthropocentric interpretation of reality. Also, thanks to the contribution of new technologies, the counterintuitive condition of many natural phenomena has become evident (Wolpert 2000). The extended dimension of somatization, as a cognitive ecosystem characterized by multidirectional interaction flows, helps us realize that a change in the extension field produces existential transformations, namely that technological interfaces produce the emergence of new forms of subjectivity.

In the humanist interpretation, the spirit's verticalization is possible through a process of progressive ascent away from the body, which somehow recalls the service modules of space rockets. A core is identified which becomes a cognitive center that directs an individual's existence (Kant 2021), so that everything around it is subject to this entity's will. Yet the more we strive to locate this entity, the more it vanishes from us. It becomes natural to assume that the cultural dimension and technological equipment are nothing but protective shells, operational tools. We could dub them "resources," just like our hands, heart, stomach and all the somatic apparatus that we carry along, like a light backpack when we are young, and a heavy burden in old age (Agamben 2015). In the humanist vision technology is not somatized because its interpretative model is not based on a *continuum* but on disjunction: the spirit is not the body, and the body is something very different from technology. We tend to assume that the essential core of the human remains untouched, or only just marginally affected by the changes in the cognitive extension that comes through the body. This is, however, a serious mistake.

Cognitive extension is yet another descriptive application of an *eco-ontological* view of subjectivity: it denies ontocentrism, the presumption that the essence of Being can be identified. When I argue that posthumanist philosophy is somehow a form of philosophical ecology, I mean to refer to the interpretative tools provided by complexity theory (Deutsch 1998) and by the theoretical principles of evolutionism (Mayr 1982), which allow us to see predicates as relational emergences rather than the emanation of an essence. If we think of the brain as the city center, our desires, emotions, dreams, gratification, sense of fulfillment or appetencies become formalized in the periphery. Instead, we should think of bodily functions like meteorological conditions: every event happening in the body produces decisions, opinions, and plans but also blood congestion and metabolic differentials. There is no control room. Our mood depends largely on the intestinal microfauna, basically on the bacterial populations we host, and 90 percent of our body cells do not have a human genome. This should help clarify the concept of cognitive extension: it is a network of interactions that may change in shape and amplitude (Mulgan 2018).

The ecosystem-body is a moment in space and time, characterized by strong field interactions that produce effects of mutual dependence and emergence. Our subjectivity is a phenomenon that wonderfully originates from all this and lasts for a certain period of time. To have an ecological notion of the body means, therefore, not only that we should not be looking for a center or an essence, but also that we should understand the functional characteristics of the field. This does not mean to remain imprisoned in a predefined and permanent cocoon but to change over time, include different spaces and be able to create *existential syncytia*, so that subjectivity may include multiple bodies and cross time through bodies. Somatization is, therefore, a quality of the body. This body is endowed with properties that are: (i) inclusive: it extends beyond the skin; (ii) comprehensive: it can create existential assemblies; it becomes an extensive-self that embraces multiple othernesses; and (iii) metamorphic: it modifies its internal configuration and the emergence of identity accordingly (Belk 1988).

Once more, it is useful to underline the difference between the humanist and the posthumanist approach. For the former, the technological apparatus is similar to a fractal or to multiple shells that enshrine the body (O'Connell 2018), or to a container for the spirit. Posthumanism, on the other hand, rejects this disjunctive view of the body as a prison of the soul that transforms the cultural dimension into a superstructure. For posthumanism, the subject is an expression of somatization, a process in the making that includes all its dimensions along with its acquired references. It is impossible to separate the subject from somatic physiology (Francis 2015) and the conjugations produced by technology. I would argue that, in general, it is impossible to separate the subject from all its field references. We are our body as much as we are the technology with which we relate to reality. It is unthinkable to discuss the human spirit and disregard the phylogenetic, physiological and hybridizing character implicit in the subject's worldly and mundane condition. What dissolves in the posthumanist vision is the *body structure*. In other words, it is the body conceived as conforming to a hierarchical structure, static, organized and confined within clearly definable boundaries.

*Being-a-body* can be disturbing for someone who comes from a cultural climate that has always projected an individual's existential aspirations elsewhere. Yet I believe that it can also be a liberating experience because the *condition of being among bodies* puts an end to the sense of loneliness that permeates the image of a spirit imprisoned in serial containers, like Matryoshka dolls. For posthumanism, the fragility and transience of *Being-a-body* are also conditions of existential fulfillment. Transhumanism, on the contrary, considers these characteristics diseases that are curable through technology. It is therefore looking for antidotes to them, thus betraying its profound humanist background (Savulescu and Bostrom 2009). This ferment of somatic participation is a celebration of life. Because it means openness toward others, life becomes richer when it is donated as a gift. It would be meaningless in a solipsistic dimension. Our body is the most evident expression of this self-giving, no matter whether we understand it or not, busy as we are, looking for a self-referential meaning. In whatever form life takes, being born is already a gift because it implies a legacy accumulated over the course of different geological eras. Our lifetime is valuable precisely because it is limited. If it were not so, we would not even have the perception of time.

Reverting to the body, whose vulnerability may inspire the noblest of feelings, means acknowledging its characteristics and bringing them to the fore. We have a certain period of time to give ourselves to the world; a welcoming impulse confers meaning to our lives without the need of transcendence or antidotes (Negri 2020). Experiencing the body as a flywheel of contacts and being able to extend it beyond the surface of our skin already gives us solace. Posthumanist philosophy returns us to corporeity, but not in a hedonistic sense. The latter is the flip side of transcendence as it presupposes individual loneliness. Posthumanist philosophy brings us back to corporeity in a participatory sense. We should not believe that happiness is about appropriation and that we are driven by attractions that are outside us. The body is fullness and exuberance, a profusion of energy that aims to flood the world. Therefore, desire is never the consequence of scarcity or deficiency; rather, it is the overflowing expression of an inner energy. To refer motives back to the somatic dimension means avoiding the trap and the misleading idea of being in credit for something: this would only invite false claims leading to dangerous illusions that hamper self-donation.

## TWENTY-FIRST THESIS

Posthumanism profoundly changes the concept of identity from the static image of the person as a mask that makes itself recognizable through morphological continuity to a mutant identity based on proxemic extension.

The body of each and every one of us is the fruit of a phylogenetic gift: it sets up our general structure through the inheritance of genetic and epigenetic traits and through ontogenetic events that begin to occur from the very early moments after conception (Gould 1977). Ontogenesis produces a somatic history, which, as we know today, affects many aspects of the orientation and differential development of organs. It is a sequence of events that are registered in the body, such as its particular synaptic wiring, the endocrine setup and the immune structure. In posthumanism, the body is, therefore, a fractal of stories that size each other up, and subjectivity seeks to conform to its own becoming, to its historical dimension (Eagleman 2020). For posthumanism, becoming—the dimension of transience and temporariness—is the most important interpretative key of the new concept of identity.

Posthumanism is, therefore, also characterized by the metamorphosis of the identity principle, which does not lead to a collapse of the concept but to its dialogical transformation. In posthumanism, identity is neither oppositional nor disjunctive. It is a *proxemic moment*, namely a way of opening and presenting oneself to the world. Identity is thus the fruit of a temporary and dialectic—hence historical—synthesis, which hybridizes a subject's previous identity with its opening to otherness. For posthumanism, identity has a historical and relational character, it is the synthesis of the subject's hybridizing experiences. In other words, identity results from what and whom a subject used to be, from how it used to feel and from what-whom it used to relate to as well as from ongoing relationships (Sloterdijk 2011). "Historical" means that identity is both the *outcome-of* becoming and *openness-toward* becoming. Identity is never pure because it is not an essence but it is the result of a *contaminating process*. We speak of a proxemic identity moment because the subject does not conform to itself; instead, it reveals a disposition toward relationships. It is an identity that oozes of otherness and expresses itself through otherness, so the dialectics between these two terms are purely historical; it is an encounter event.

The body's events tell us about physiological transformations within the flesh (Deleuze and Guattari 1997). They also have an ontological meaning. All of them reveal how fragile and dependent on external factors the subject is during the course of its existence. Individual life resembles a series of gradual openings to the world, rather than appropriations: they should be interpreted as gifts and donations that trigger processes of decentralization (De Gainza 2014). Every relationship enlarges a subject's proxemic field by removing it from its previous condition and making its identity more and more convivial. An individual's biographical experience thus becomes an itinerary of deindividuation or participative growth. Yet as an expression of events that concern and happen in the body, experience cannot be based on itself. In other words, we cannot base experience on experience itself. Experience is received by the body because it relies on an ontological a priori, the *Having-been-there-before*. While in the humanist vision, based on the ontological autonomy of *anthropos*, becoming is a progressive detachment from the body's reasons, in the posthumanist vision, it is the realization of the body's reasons, conceived as a gift to the world (Spinoza 1951). In the transhumanist vision, the body is a gravitational center. It is for this reason that transhumanism seems to be a continuation, in a technological disguise, of the twentieth-century hedonism that followed the collapse of the great immanent ideologies (Del Noce 2012). In a hedonistic and narcissistic perspective, the body is supposed to adhere to an aesthetic model that denies its specificity and differences. Most of all, however, it denies the concept of change. Hence, the body's events are experienced as disfiguring. Posthumanism, on the contrary, is a liberation of the body because it acknowledges the ontological character of *Being-a-body*: openness to the world and self-giving. The body frees itself, in the sense that it accepts its relational nature and what sustains the subject's existence.

A body that extends and appropriates the references with which it establishes relationships, and that does not remain stable and confined within the limits of its phylogenetic architecture, is necessarily a metastable entity (Stiegler 2013). Its metamorphosis is not radical and situated like the Kafkaesque one, though, but it is continuously being reworked. Besides undergoing small transformations, this entity also shows some persistent elements, which, however, change over time. Therefore, there are no common denominators other than a historical recognizability. Metastability defines its continuity. What I recognize as related to identity is the consistent becoming of the subject, even though between the points of beginning and end, all the individual's predicates might have changed. But metastability is not only temporal; it is rather something mutable, which defines the identity space as a relationship space that goes beyond the body's liminal profile and turns into a sharing space. Identity is thus no longer the expression of an individual but the manifestation of a *convivium*, of shared proxemics.

The perception of the Self, in fact, is mediated by the channels that give it access to the world. They build a proxemic image that progressively enlarges by gradually including new perceptions. This self-image extends beyond the skin thanks to the support of remote transducers that project the body beyond its anatomic limits. Remote transducers translate into recognizable, self-related images also images related to prosthetic structures. While identity does not refer to an essential nucleus, it is also not contained under the skin or the sensory apparatuses. We speak of a proxemic identity, which consists of gradients of self-perception. We realize that the proxemic bubbles (Hall 1990) that define identity do not pertain to the individual, because identity spaces are differently shared with whomever surrounds us. This means that the sense of Self is a *community sense*, as it is not exclusive to the individual but shared with the group (Esposito 2006). Proxemic identity is, therefore, relational and continuously mutable, it is a dynamic entity whose shape changes like a cloud in the sky, where certain aspects tend to be more stable than others. This is what I mean by metastable and shared identity.

Thanks to the contribution of technologies that provide constant access to connective dynamics, the extensive dimension of identity has become more evident. Due to the human's strong copulative nature, it has actually always been like this. In the time of the Indo-European revolution, the Steppes peoples, who were the protagonists of the great migrations on the Eurasian continent, were, in all respects, centaurs: their sense of self extended to the horse, they also perceived themselves through equine muscularity; that is, equine muscularity contributed to their self-perception (Diamond 1997). The same is true of the Arctic hunters of the Paleolithic Period, who were completely hybridized with their dogs and both of whom projected onto, and extended toward, the megafauna of the Pleistocene (Shipman 2017). This extension continued with the transformations of the Neolithic Period: the peasant was not only the man who plowed the soil for sowing, but also the macroorganism that incorporated it along with the oxen and the plow. They formed an extended identity where each of them reciprocated the other's action. In the same way, a predator is also the prey and vice versa: they end up by mirroring each other, like the oak's crown mirrors its environment and the environment mirrors the oak.

The disjunctive tendency is again the error of the humanist perspective, which conceives of identity as an introflexion, a retrospective adherence to a primordial form that represents the principle of individuation (*principio individuationis*). This approach predates humanism, but it is in this cultural climate that it gained traction. The disjunctive and oppositional identity is instrumental to the emergence of the human as a vertical and dominant entity, withdrawn within itself, autonomous and self-founded. Unfortunately, its distortive and misleading effect can be observed in today's solipsistic and narcissistic drifts (Lasch 1979). Because of their interactive character, the creation of particular interfaces, such as cybernetics, informatics, and biotechnology, makes this aspect obvious. The posthumanist proposal moves in two directions: (i) like the Copernican revolution, it is an interpretative model that does not introduce a new phenomenon but changes our way of reading an event; and (ii) it is a project that imagines and defines the possibility of a different humanity, one that lives in harmony with the world and enjoys a greater inner balance.

Unlike the Cartesian *cogito*, this metamorphic identity cannot remain true to itself or protect its private space because everything is continuously being negotiated communally. Yet we don't need to be frightened. This identity is something that should induce us to revise our notion of coherence (Arecchi 2008). Coherence does not mean that we should hold on to our ideas, but that we should accept the consequences of our dialogues, both when we agree and when we do not. Eco-ontology is dialectical, just like an ecosystem that builds its overall synthesis through a matrix that combines different interests, sometimes even opposing ones. An ecosystem is a metastable entity: it is recognizable not because it remains identical over time, but because its evolutionary plot creates a story. The historical condition of identity might seem inconsistent because it tends to change. However, this is an interpretative mistake. We are inconsistent only if we do not change after our ideas have been refuted; in other words, if, in order to maintain a chimeric loyalty to ourselves, we interrupt the flow of dialogue or accept some external entity's authority. Freedom as participation means, first of all, accepting to be part of a community. Hence, we avoid solipsism, that is, existential echolalia, as well as subjugation to a dominant principle. This is the reason why it is correct to speak of a critical identity.

What twentieth-century thought has put in front of our eyes is a progressively fading image of the human, an image drawn on sand and eroded by successive storms (Foucault 1994). The clear-cut notion of identity as a single entity, different from all the others, with solid predicates and "special" has collapsed. This process is irreversible, and we have to accept it. We have received many blows in a relatively short time. Paleoanthropological discoveries, for example, have revealed that multiple human species were populating the Earth up to 50,000 years ago. This discovery caused a far-reaching perspective shift in relation to the idea of human specialness and uniqueness, comparable to the reversal of perspective brought about by cognitive ethology with the discovery of intellectual and cultural qualities in other species. After the Darwinian revolution, the already fragile image of the human has received many other blows. Rather than Norbert Wiener or Alan Turing, it is Charles Darwin who should be considered the forerunner of posthumanism, because he was the first to establish scientifically that there is no discontinuity between humans and nonhumans. After the Darwinian tsunami, the great progress of twentieth-century biology further laid the foundations of the posthumanist transformation. It disclosed the transversality of the genetic domain that eventually wiped out the essentialism of predicates. We cannot overlook the offshoots of these paradigmatic shifts.

Shared identity is, therefore, first of all, a different way of considering subjectivity and the Self, who suddenly feels free to experience new existential dimensions. The interpretive revolution ignited by the findings in biology and by the opening of identity borders, has promoted a new cultural climate. These findings include, for example: the acknowledgment that we share predicative attributes with animal otherness; the organism's morphogenesis as an entity on which it is possible to intervene; and new techniques in procreation and genetic engineering, genome editing and biosynthesis practices. This is a zeitgeist that creatively transforms what used to be the preservation of a primordial form. Both reconstructive surgery and cosmetic surgery—the latter of which came from the former—are linked to a homologated model of the human image. On the other hand, morphogenetic creativity produces new body aesthetics that are closer to concepts such as fashion clothing or makeup—hence, they are transient. Subject to the collective influence of fashion, in this model of aesthetics form is something contingent. It is not a person's attribute or something stable. While the concept of person-mask represents continuity and self-presentation, expressing our wish to be recognized and to appear in a particular way, the mutant identity (Balibar and Morfino 2014) feels and experiences the body in the flow of different situations. It is transient because it changes along with the relational conditions that make up subjectivity. In other words, the open and free interpretation of the human image that derives from its body's horizontal transformation produces a new experience, which affects how we experience our corporeity.

We could also argue that posthumanism is a revolution of the flesh (Zaltieri 2009), seen as free morphopoietic productivity, an emancipation from the constraints of bodily stability. In this proposal, the body's events become ontopoietic experiments based on processes of projection into, and reception of, the world. The notion of a proxemic moment enhances the body's positional meaning: its organization cannot be viewed as autopoietic or as liquid or as dependent on external determinants. It is a revolution of the flesh that is not a denial of the body but an acknowledgement of its dialogical character, of its true ontological constitution made up of openness and becoming. Every single bodily event produces a metamorphosis in the matrix responsible for its proxemic dynamics; it defines new somatic architectures and adjusts our relationships accordingly. The body's dynamism and plasticity, its being permeable to the contribution of external reorganization, makes the flesh an *open building site* that constantly renegotiates its morphological plans. The body is made for dialogue and is transformed through it. In the posthumanist view, each of the organizational levels that are present at a given moment are springboards of somatization, a dialogical becoming that enables the body to continuously go beyond itself.

There is an ontological and aesthetic difference between the creative and relational metastability of the posthumanist body and the plastic and impervious staticity of the humanist somatic dimension. Such a difference cannot be trivialized by technological makeup and masks. However embellished by cybernetic exoskeletons, transhumanist images usually conform to a sort of humanist mannerism (More 2013). Whether represented by a robot, whose epidermis is replaced by a sheet metal surface, or by an asexual and angelic entity, transhumanist imagery does not deviate from the humanist canon, as becomes apparent from Renaissance representations of the body, based on perfection, purity, proportionality, and performativity (Sandberg 2013). It is a body that claims its autonomy from the world and, accordingly, imposes itself on its surrounding reality by both dominating and withdrawing from it. Identity remains static and uncontaminated, totally impervious to a world that is sublimated by it and permits its ascent.

The posthuman body, on the other hand, is a temporary form in continuous transition, precisely because it is continuously in a relationship with the world. The tendency to enhance differences and proxemic temporariness has become stronger and stronger over the last few decades, together (and in contradiction) with an emphasis on the humanist canon of the body's homologation and the attempt to deny the process of somatic becoming or aging. Experimentations on the body, such as tattooing, scarification, and piercing, became increasingly popular during the second half of the twentieth century and were the forerunners of this cultural atmosphere. The skin surface turns into a canvas on which to write one's own story, assuming that identity has more to do with a historically configured relational milieu than with adherence to an essence. This consciousness grew in the subject during the transition of decades between the two centuries.

Mutant identity might seem to be a vindication of somatic freedom and a rebellion against what is predetermined; it is a subjectivity that claims emancipation and full ownership over its somatic dimension. It is obviously not just a matter of appearance or aesthetics, because it also refers to the subject's sexuality and performativity. However, I believe this interpretation to be the retrospective expression of a humanistic model. Mutant identity is something different from appropriation or freedom space (Caronia 1996). Arguably, it is exactly the opposite: it is expropriative, a way of making our body adhere better to the flow of the relationships we want to establish. It is a body that harbors multiple relationships and that the subject tries to adjust more and more to his/her own feelings and experiences.

## TWENTY-SECOND THESIS

Posthumanism starts from a copulative conception of the human, and therefore rejects the principle of deficiency as the expressive driving force of the human, considering agentivity to be the consequence of internal predicative exuberance.

The concept of the action drive—what moves the action—is among the most important innovations of posthumanism along with the idea that subjectivity is shaped by experience, rather than given. Let me clarify these two points and begin with an analysis of the notion of subject. The Cartesian *cogito* provides a rationalistic explanation (Laporte 1997) of the subject's attitude toward itself and the world. This tradition is what mainly supports the humanist approach with its notion of human self-foundation and anthropoplasty. What stands out in the thoughts of Arthur Schopenhauer (Schopenhauer 2010) and Friedrich Nietzsche (Nietzsche 2017) instead is that the drive of desire is ultimately a projection of perspective will. Regardless of how will is differently interpreted by the two authors—as an ensnaring appetency for Schopenhauer and as a creative and vital force for Nietzsche—posthumanism is much closer to this line of interpretation of agentivity than to that of rationalism. This does not mean that posthumanism derives from some form of ontological irrationalism. It means that subjectivity emerges from the vital flow of the desiring entity that I have called *Having-been-there-before*, so that agentivity is not based on an a priori subject, who has desires. In this sense, posthumanism is a philosophy of becoming, emergence, internal exuberance, and, above all, the Dionysian indistinctness that encompasses all the vital forces that lead to subjectivity.

Instead of postulating an a priori subject who gains self-consciousness through *cogito*—or a free subject who, by being free, can cogitate (Lequyer 1998) or, yet again, a subject who states its presence by *cogito se cogitare* (Heidegger 1991)—we have to think of subjectivity as something that forms gradually. As an entity that can think about itself, the subject emerges from a desiring drive that, like a verb, conjugates its phylogenetic legacy by means of copulative experiences of the world. We are not born subjects, but we become them by opening up to the world. What should also be clear is that we do not become subjects because we become aware of a Self, but because, by relating to the world, we unfold a Self. As it lacks a unified organization, this openness is not exactly a will or appetency. It is rather a tensional flux that expresses the concept I have baptized: *Having-been-there-before*. It means finding the world again through previously experienced contents, that is, through experiences that are neither firsthand nor specifically human. The drive of desire is not a will imprinted on the world but a yearning for the world, a desire to join the world again. It resembles a letting go of the Self to the flow of a Dionysian current (Nietzsche 1995), a flow that is vital but also choral and unspecific, that is, not enshrined in an essence. The internal exuberance that characterizes our first unfolding in the world, this experiential dawn, cannot be attributed to an entity endowed with a precise and defined subjectivity. It is rather like a feeling of surprise, wonder, fascination, a feeling that engages us and a force that drags us along. When it is so interpreted, considered something different from appetency or will, the force of desire produces copulative agentivity. It is from here that self-reflection and self-ownership—thinking and awareness—emerge. Subjectivity is not given a priori, but it is the fruit of agentivity, the result of an openness to the world that allows the development of a reflexive Self. As the shape and structure of a tree do not precede its growth and are not the mere algorithmic unfolding of internal contents, so subjectivity is a process and not a foundation. Through its dialogue with the world, the Self conjugates the drive that pulls it toward the world and acquires contents of its own.

The core ontological issue of posthumanist thought is the subject's hybrid dimension. The relational nature of the subject's predicates originates here. This is why posthumanism can be considered a philosophy of *relationship constituency*. Although implicit and not always expressed coherently, hybridization is a founding rather than merely a potentiating or juxtapositional moment. Indeed, we speak of hybridization as an ontopoietic emergence. This definition supplants the idea that the world is a home or container as well as the instrumentalization of otherness. The subject emerges through its relationship with the world (Simondon 2020), and adopts a perspective dimension. It perceives the world in a particular way and confers a particular meaning to it, which does not preexist the hybridizing encounter. Our perspective over reality is also always a construction of meaning. When we move around the world, each of us notices things that make sense and are interesting to us, and that are different from what others notice. Subjectivity is ultimately making sense of the world. Precisely because of this process, the subject always comes into being through relationships, that is, through a web of bonds. The subject, however, is not the *ex nihilo* construction of experience because experience itself stems from the desiring drive of the *Having-been-there-before*. The *Having-been-there-before* meets the world and resonates with it, finding reasons to open up. This copulative pull—this a priori enthusiasm toward the world—finds its coherence again because it is the outcome of an existential flow molded by the world (Lorenz 1973), and not because it has already been a subject in the world. The subject is also the a priori expression of the world and the a posteriori result of its relationship with it. Along with the development of emotional bonds with otherness, the subject's existential itinerary is also a big web of introjected relationships. It is an affective entity in Spinozian terms (Deleuze 2001) that, by being interested, becomes cognitive and can make sense of the world: it expresses an ontological dimension that is never disjointed but always ecological. The subjective condition is always diachronic. Hence, we can argue that the subject lives in the here and now, because it emerges *from* the moment rather than being immersed *in* the moment. The present is therefore a construction the subject makes by *being always elsewhere*, namely living in an extended moment that is as strongly rooted in the past as it is projected into the future. This perspectivism can be defined within a *cogito* that, however, is never abstraction but always connection, both when it reflects upon itself and when it reflects upon the world. Subjectivity is not consciousness but that which makes consciousness possible, including dreams, unconscious imagination, ecstasy, enthusiasm and psychic alteration. But it is precisely within the subject's perspective plurality and polyvocal condition that the proxemic identity emerges, an identity that is simultaneously recognizable and mutable.

It is now essential to underline the difference in how humanism and posthumanism interpret desire. I believe this to be the basis of the existential force of posthumanism. As previously argued, humanist thought revolves around the principle of human deficiency, which is a secular version of the Augustinian concept of *natura lapsa* (Augustine of Hippo 1913). Hence the idea that: (i) culture compensates for an original deficiency; and (ii) desire expresses this deficiency. If becoming is to be understood as an open project, posthumanism cannot help questioning the principle of human deficiency. To do this, we must show the contradiction behind the idea of the human as a demiurge and a motor on the one hand, and on the other, the need to find reasons for action outside the self. The deficiency principle is convenient because, by creating a mythopoetic and self-founding image of the human (Olien 1978), it helps conceal the contradictions inherent in these positions. In the first case, the motive of agentivity is the adaptive need caused by the Epimethean deficit. In the second, it is the supposed attractiveness (desirability) of an external entity. By denying the existence of an inherent expressive drive and relying on the artifice of compensation, the deficiency principle elegantly avoids considering the existence of an a priori condition regulated by implicit directives. With regard to desire, the deficiency principle is almost tautological: it assumes that we can be aware of something beyond the here and now—hence *de-sidera*—from within a condition of deficiency. On the other hand, the myth of biological incompleteness, or ranklessness, prises humans apart from animals. This is because animals are complete, hence they have no freedom of action or agentivity. For humanist philosophy, therefore, the double configuration of deficiency is a formidable device with which to continue applying the precept of separation and disjunction. For posthumanism, on the contrary, action is motivated by exuberance, an exuberance that features in life and transforms itself into copulative openness, amazement, and the desire to embrace life.

The core challenge to go beyond humanist ontology is, in my opinion, to revert to desire as the founding principle of subjectivity and to reject the appropriative interpretation, which offers a teleological explanation based on the idea that the object-result is desirable because it is missing. In fact, desire is an exuberant drive that drags the subject into the world. Desire can also mislead us and create the impression that we are missing something. But we cannot account for desire as the lack or nostalgia of something. Copulative events are only apparently appropriative. They are actually predicative or ontopoietic actions. When we speak of hybridization, we must not think that the subject is empowered but that it undergoes a metamorphosis (Marchesini 2021), a change in its inner organizational guidelines. This implies, for example, that not all technopoietic events are triggered by the need to compensate for a lack. Copulative events are projections beyond the existing condition. They are a push toward renewal that is justified by exuberance. Yet hybridization is not something technical or specifically human. It is the very foundation of life. It is the tendency to create connections and dependencies, that is, phenomenologies of life that depend on the world. And agentivity is the desire to be part of something, an opening up to the world and a willingness to hybridize. Agentivity is centrifugal (a gift and dependency) and not centripetal (the appropriation of something). The principle of deficiency, on the contrary, views agentivity as compensation that, by separating the subject from the world through the concept of the limit (Plessner 2019), makes the human autonomous and independent. The principle of desire moves in the opposite direction and views ontopoiesis as a process geared toward heteronomy.

Heteronomy can be described as the inner tendency to create annotations and glosses that refer to the world as if it were a text—the gills of a fish refer to the presence of water. These annotations and glosses might also seem to be needs and, somehow, that's what they are. However, it would be a mistake to consider copula (the expression of agentivity) compensation to gain autonomy or ontological immobility. All expressions of desire push further and further toward heteronomy. We could even argue that the whole of life is Promethean, because it is based on continuous references to the world. This adding of references is the ontological expression of Brentano's intentionality (Dennett 1996), and it cannot happen without heteronomy. The process of life is therefore brimming with exuberance, an uninterrupted becoming that creates individual subjectivities by dispersing and depleting exuberance and becoming contaminated with the world. If desire triggers agentivity, the relationship between the two (desiring and agentivity) is like that between the verb and its object complement: the expressive copula defines the object of desire. But it is not the lack of the object that triggers desire. If hybridization is an act that does not develop self-referentially but by virtue of an opening and copula, it needs a specific verb to realize itself (Lichtenberg 1989). By what actions do we address the world? It is the question that defines the verb that directs the entity outside. The answer provides object complements that outline the verb's context and situation (toward what, how, where, etc.). All encounters unfold at levels which, as well as expressing a general interest in the world, must be the concrete execution of something, such as taking care of someone or something or imitating and interpreting a phenomenon. An *anthropology of relationships*, therefore, needs to clarify the type of agentivity that develops in the encounter (Held 2006).

In order to express their vital state, all organisms must use a hosting dimensionality, an opening structure in their system, that is thermodynamic and ecological. They must also deplete the information inherited by phylogenesis. Life can only exist in a copulative sense, both in synchronic terms, as noted by Erwin Schrödinger (Schrödinger 1940) and Elya Prigogine (Prigogine and Nicolis 1977), and in diachronic terms, as a depletion of the legacy. Heteronomy is therefore the cardinal principle of life phenomenology: the drive of desire depletes the legacy by creating new bonds. With its propensity for bricolage (Jacob 2020) and niche construction, technology also follows this logic. Niche construction or a species' tendency to modify the environment in order to create its own living dimension is a clear expression of the organism's dependence, which finds accomplishment in speciation. Niche construction is also a new form of species-specific legacy. Considered from this perspective, technology is the fruit of human creativity rather than of an adaptive deficiency. It is a form of heteronomy, the development of a dependence.

Conceiving of technology as compensation and a tool of empowerment and autonomy, as contemporary hyperhumanist expressions suggest (del Val 2012), causes us to project ourselves into a future that is profoundly different from the posthumanist one. For posthumanism, *téchne* is the fruit of a creative dialogue with the world, which produces bonds and dependences. In the posthumanist view: (i) technopoiesis does compensate for an insufficiency, but it is a creative event, only just partially under rational control, which produces bonds and dependences; and (ii) technomediation does not estrange the subject from the world and, above all, it does not verticalize it. It enriches its perspectives, hence its means of connecting with the world. Let us now observe what it means to interpret creativity as a dialogical moment. It means: (i) to acknowledge that the human is attracted to the world and willing to project itself into it as well as be permeated by it; and (ii) to view the result of conjugation as creative, the fruit of a hybridization with the world. This creates a link dependence, a close bond with otherness. This is why *téchne* can ultimately be viewed as an expression of heteronomy.

If desire is exuberance, it becomes a spur to action, while copula is the undertaking of action, and dependence and bonding are its consequences. On the other hand, if we assume an *ab-origine* deficiency (Herder 1869) with a resulting compensatory view of culture, and of desire as a lack, deficiency becomes the cause of agentivity and autonomy its goal. This is the stance taken by humanism, which reads agentivity as the empowerment and elevation of a subject who tends to become more and more autonomous and detached from the world. This withdrawal into a solipsism based on self-sufficiency and autopoiesis shows a strong continuity with the transhumanist proposal. Transhumanism presents itself as a philosophy that liberates the human from all the limitations and constraints of its organismic condition (Kurzweil 2005). Apart from its picturesque aspects, the transhumanist illusion of a postorganic and immortal existence is the same ontological—and, consequently, anthropological—interpretation of the motivation behind human action as we find in humanism: there is a gap to fill. This is the reason why I see no similarity whatsoever between transhumanism and posthumanism. Quite the contrary, as posthumanism inverts the mechanism that accounts for human action; while transhumanism sticks to the humanist view of the subject as *primum movens*, posthumanism sees human action as an exuberance that creates subjectivity by hybridization. In fact, what we find in transhumanism, albeit not clearly or coherently, is a tendency to prolong modernity's immanent eschatology simply by replacing spiritual emancipation with technological emancipation (Valentini, 2015). Exuberance stands in opposition to deficiency. The former views the drive of desire as an opening toward copulative events, which: (i) depletes past contents; and (ii) creates an itinerary of unique unpredictable and indeterministic renewal. Exuberance brings an inner motive, which is compensatory but hybridizing. This motive enriches our connections and bonds with the world by making us dependent on it.

Since the difference between the two visions is instrumental to an understanding of the posthumanist proposal, I would like to emphasize that: (i) for both the humanist and the transhumanist interpretations, an a priori deficiency lies at the basis of the ontopoietic drive. This deficiency creates constraints from which the subject wants to free itself by an act of both compensation and separation; (ii) on the contrary, in the posthumanist interpretation, ontopoiesis is the consequence of a brimming exuberance that, like a river in full flow, pushes the subject into the world to build deeper and more articulated relationships with it. For posthumanism, the ontopoietic movement is neither vertical nor disjunctive, but horizontal and connective. The thinking subject who becomes self-aware is an entity already equipped by the world. It is not a cogitating a priori who is free by the grace of God. Its perspective on the world is, therefore, polyvocal. It originates from different experiences, not always consistent with each other, and is continuously in a process of becoming. The rational effort that is required, therefore, must consist of an attempt to find syntheses, albeit transient ones, in this multiplicity of perspectives. Creativity must avail itself of rationality, yet it remains the fruit of the desiring condition, hence of a copulative agentivity that produces heteronomies. Creativity can neither be self-sufficient—based on itself—nor produce self-sufficiency.

If we take the discussion to an anthropological level and illustrate how the drive of desire has boosted the development of creativity and led to the emergence of culture and *téchne*, we need to stress how creativity is strictly dependent on relationships with the world and on experience. The outside world exercises an irresistible fascination on the human, who needs dialogical referents. This relational drive does not seek compensation but accessibility in the world. In other words, it comes into the world to find fulfillment through self-expression and dissipation. Copula is therefore not a clear-cut project, but the consequence of our openness to the world. We could argue that exuberance and openness are ultimately the same, whereas hybridization is the natural consequence of copulative agentivity. The world does not simply attract but produces *thaumazein* (Berti 2007), namely wonderment and thrills, involvement and participation, all of which transform desire into action. The accessibility that the world offers is not neutral or undifferentiated but correlated to the force of desire: the cat does not observe the world in generic terms but consistently with its desires. The entity is not attracted by the world but its exuberance returns to the world and, by so doing, creates a distinctive subjectivity. The subject is, therefore, the fruit of the principle of desire, whereby the *Having-been-there-before* continuously renews itself by shaping always new subjectivities. The copulative experience is a centrifugal movement toward the world. We could almost describe it as the need to explode in the world outside, to conjugate ourselves with what surrounds us, to fulfill our potential for taking action and convert our projections into action.

An anthropology of relationships is not like traditional anthropology: it does not claim to create the human *iuxta propria principia*. Posthumanism confers to the ontological dimension a sort of theorem of explicative incompleteness (Hofstadter 1979). It postulates that the human cannot be explained by self-investigation and by studying the human from within. In this sense, posthumanism is a philosophy of relationships. In posthumanism, the hybrid condition is the constitutive principle of Being whose predicates are created through relationships. Experience produces bonds because it tends to address otherness and affect well-rooted existential contexts. Bonds are therefore the basis of our dedication to the world. The latter is the fruit of desire and of a feeling of being in harmony with otherness. The *emotional bond* is the true principle of subjectivity: we may also experience it as a deficiency (albeit as a consequence and not as a cause of desire), but it is actually the most explicit expression of life phenomenology: attachment to the world. To interpret deficiency as the justification for our actions, therefore, would be a mistake. Desire is actually the consequence of experience and action. Experience cannot be based on itself: it is the *Having-been-there-before* that spurs engagement, that state of wonder and tremor that triggers the copulative movement. The principle of desire is therefore a force of attraction, whose field of action is our encounter with the world. Rather than a self-referred and disjointed action, however, this encounter unfolds in the space of hybridization. The stronger the agentiviy, the greater the need for an expressive field, and consequently, the tendency to connect with external entities and develop heteronomy. If we interpret desire as appetency and an urge for possession, it produces a dependence that causes suffering and that we want to eliminate. If, on the contrary, we consider desire as a copulative moment that creates an emotional bond because we give ourselves to the world, dependence provides existential relief.

## *Chapter 7*

# A New Culture of Téchne

Having a different idea of the design process as a starting point, posthumanism reviews the human relationship with *téchne* and offers a new interpretation of the relationship between the body and technology (Capucci 1994). First of all, imagination is no longer a solipsistic and autarkic moment, but one of a profound relationship with nature, a condition where the inventor draws inspiration from its ecstatic relationship with the world. Once more, we question the solitary and emanative view of the human: this is in order to reiterate the importance of a dialogue with nature in all forms of human expression. Furthermore, the body is considered an entity that is not simply empowered but transformed by technology. Technology is somatized and, by blending with the body, it redefines its organization, especially from a functional point of view (Andrieu 1993). Hence, we speak of *technophysiology*, because the device operates like a virus, modifying the metabolism of the cell-body.

### TWENTY-THIRD THESIS

Posthumanism is a reinterpretation of the concept of *téchne*, no longer as a solipsistic moment of human ingenuity, but as a revelation or relational process. This process is not limited to human enhancement without changing the human perimeter but produces hybridization, that is, the emergence of new predicates.

The posthumanist view of technology rests upon the idea that the entity is a hybrid, an ontopoietic becoming that derives from a constitutive relationship. Therefore, posthumanism sets up a completely new reasoning with regard to the following: (i) the emergence of the technical event, *technopoiesis*; (ii) the experience of *technomediation*, a new relationship with the world; and (iii) the consequences of *technical somatization*, changes in our existential dimension. It is impossible to understand this profound paradigmatic change if we just focus on the impact of new technologies without changing our interpretative model of technology (Galimberti 1999). As previously suggested, new technologies have highlighted the hybrid dimension of the human predicate and stimulated a new reflection on the overall meaning of *téchne,* also in retrospective terms. In fact, the chipped stone also produces hybridization, and, in more general terms, niche theory itself is a stimulus to reconsider life phenomenology. The issue of technology is intimately linked to our model of the human condition, beginning with the very notion of human nature. The extent to which the various inflections of the humanist tradition have contributed to building a die-hard view of the human is the starting point of the arguments that follow (Hughes 2004). This is the posthumanist challenge: to create a new way of thinking about technology by beginning with a new way of thinking about the human condition. In order to do this, however, we must perforce revise our concept of human nature, which links to how we perceive nature. As long as the phenomenon of life is viewed as pure extension, without creativity and freedom, as incapable of transcending its status, we will never be able to consider human nature as anything other than inadequate and deficient.

In order to understand posthumanist thought we need to exit dualism. This does not mean that we have to convert to biological reductionism, though. In fact, it means we need to profoundly transform our way of looking at life phenomenology. Today, Bergson's proposal of a creative evolution in the living, of an experimental vitality that is free and unpredictable, finds new epistemological foundations (Marchesini 2022). Let us begin with how the applications of biosynthesis and *extended synthesis* in evolutionary theory (Pigliucci and Müller 2010) have modified our interpretation of nature over the last century. Thanks to these contributions our picture of the living differs significantly from the static image of the humanist tradition, which was still somehow hinging on essentialism and determinism. Today, all the living is viewed as an entity that creates its existential dimension by exercising action upon the world: building alliances, leaning on the outside world and co-constituting itself through otherness (Douglas 2014). Nature is multiple and creative: since different inheritance matrices—genes, proteins, and traditions—allow it to introject the contents of previous experiences, it can also go beyond what pre-constituted it. Ecology reveals a living world that cannot be eradicated from its ecosystemic exchanges, just as physiology occurs via a metabolism based on the open system structure of life phenomenology. An entity's openness toward external contribution makes *hybridization* inevitable. Nature has a plastic dimension and a copulative directrix: while setting up tendencies, it is not deterministic and cannot establish predicative consequences. These are always the result of relationships and cannot be assumed.

A system of thought can be said to be *post*humanist only if it pursues and develops a hybrid ontology and considers predicates relational outcomes (Simondon 2020). A philosophy is *post*humanist if it goes beyond the principle of an original human deficiency that needs the support of external crutches and yearns for autonomy. It can be called *post*humanist if it can get rid of the idea that a human essence may be clothed in technology without being modified by it. It can be called *post*humanist if it can strip itself of the need for a disjunctive verticalization through processes of purification in order to secure its *dignitas*. As long as we hold on to the idea of human empowerment or of the transcendence provided by technology (Mumford 1934), we stay inside a coherently humanist ontological framework: it is not posthumanist but it is, at best, hyper-humanist. Posthumanism demands a profound reflection upon the *concept of hybridization* by putting the ontological question of life and human nature in the foreground. The hybrid is the fruit of a predicative act, an encounter that upsets the inherent ontological plan. In other words, the hybrid requires: (i) that we question the classic interpretation of inherence, of human nature; and (ii) that we define a predicative process that does not simply extend or strengthen inherences, but transforms their existential condition. Hybridization is not a garment or a home for the *Being*, and it is not compensation or dispensation, either: it is an act of contamination that creates new predicates. Hybridization is, therefore, an act initiated by a desiring drive that produces opening and copula.

Hybridization is ultimately a creative event, originating from the exuberance of human nature and based on its need to build *constitutive relationships*. The drive of desire inherent in life phenomenology, the generative vitalism of the living, is the same that stimulates an uninterrupted ascent of the mountain of the real by relying upon previous hybridizations, namely events that have already been introjected. The leaning on the outside that we have called "heteronomy" is an existential dimension for which hybridization is an implicit *status* and becoming the implicit consequence. The desiring condition produces an intrinsic teleology of the living (Spinoza 1951). This should not be confused with the idea of an intelligent evolutionary project or a yearning for progress. However, it should not be reduced to teleonomy either (Monod 1970). The desiring condition engages the subject more and more in its own singularity. But the creative impulse is not the consequence of a deficiency, but the eccentric exuberance resulting from hybridization. All the living rests upon a legacy that eschews a synchronic explanation as well as upon a projection that goes beyond legacy itself. This is why, in ethology, Niko Tinbergen defined *four explanatory matrices* (Tinbergen 1963). For example, in order to explain the migration trend in swallows, we cannot just refer to the physiological changes occurring in their bodies. We need to refer to the history of the species' evolution and provide a diachronic explanation. Ultimately, once we acknowledge the entity's hybrid condition, our way of understanding life phenomenology radically changes.

The philosophical reflection introduced by posthumanism differs significantly from the tradition of philosophical anthropology, imprisoned as it still is in an attempt to trace human self-foundation (Scheler 1961). This is for three main reasons: (i) firstly, because posthumanism considers cultural making the result of a desiring drive; (ii) secondly, because it views the cultural outcome as an outward movement (decentralization), and not as compensation for adaptive deficiencies, (culture shows dependence, not autonomy); and (iii) thirdly, because anthropopoiesis is considered a hybridizing event rather than emanation. This is why it is impossible to infer predicative consequences by studying the human from within. When posthumanism posits that the hybrid is an existential condition, these paradigmatic shifts are taken for granted. Posthumanist philosophy cannot exist outside of an *eco-ontological idea of Being* (Marchesini 2018). Every single predicate must be considered the outcome of—and emergence from—a relationship rather than viewed as autonomous and self-referential. *Téchne* becomes the core moment of this interpretive metamorphosis, because it is the result of a relationship and it produces heteronomy or dependence. This leads to a relational view of the entity that is open to becoming: it is impossible to predict the consequences of technopoiesis and technomediation (Kelly 2011). What a conception of nature as extension risks jeopardizing is human decisional and perspective freedom, which does not happen if we consider technology the expression of its copulative drive. Not only is there a will to become *technical*, but also total freedom in the making of *téchne*. Life phenomenology must shift from the notion of an extensive to that of an intensive nature (Deleuze 2010).

In both the compensatory and potentiative views, *téchne* was seen as either the result of a self-referential imagination or the shell needed to compensate for human incompleteness. The advent of the hybrid upsets these terms and frees *téchne* from a priori confinements. *Téchne* is the realization of a copula: a human act that, through desire, builds a relationship with the world. If we realize that it is impossible to predict how dialogue and interaction can develop, it is also clear that our present needs and imagination cannot tell us anything about how our technical future will look. *Téchne* will always be a surprise, because its implementation is never self-referred, but always relational, that is to say, dependent on encounters and how they happen. Yet this development is not regulated by chance but by human specific orientations: what we have called inherences and *Having-been-there-before*.

Hybridization demands from us that we delve into the process of ontopoietic emergence that happens through otherness, rather than just assume it in general terms. In essence, I believe that posthumanist thought is called upon to formulate different criteria for philosophical anthropology. A new philosophical anthropology needs to account for the emergence of new qualities and existential purposes by envisaging the possibility of the unpredictable and relinquishing the humanist linearity and progress model. If, rather than repeating itself, nature tends to continuously transcend itself, acknowledging that human nature has specific orientations directed to the world does not mean having a scheme that anticipates the results (Prigogine and Stengers 1979). For example: saying that the human is inclined to hoard and collect does not mean knowing what it can hoard and collect. In order to build an anthropological human model that may be called posthumanist, we must translate the concept of desire as a drive into an explicit view of the process of contamination. While this model is based on inherences, it must also acknowledge that *téchne* is an event with a historical dimension. Exuberance is a push toward action—*Being-there* is not the state of contemplating reality but an immersion that transforms us. And every single action refers to history, so it can only be told and narrated, not assumed. What this means is that the relationship is constitutive, namely, it is not simply dialectical or interactive, but it makes predicates emerge. *Téchne*, therefore, is the expression, not the support, of the action. The goal of the acting subject is not to adapt, but to go beyond itself (Nietzsche 2021). Even if it draws entirely on the subject's phylogenetic legacy, *téchne* is never simply the expression of its legacy.

Human desire can be thought of as a tension that unravels through recurring verbal predicates. They are established by the human phylogenetic legacy and include: (i) mimetic tendency; (ii) collaborative vocation; (iii) epimeletic parental care; and (iv) interest in collecting and hoarding. They also define the main type of copulative agentivity: to imitate, to collaborate, to care, and to collect. Although desire does not involve a specific objective-result, it does not mean that it lacks intentionality (Dennett 1997), a reference content: to understand this, we just need to shift from the object of desire (e.g., stamps) to its verbal predicate (e.g., to collect). Desire is nonspecific. It is a tendency to access the world through recurring actions—defined by the verbal predicate—that are totally open. In a sentence such as "the cat chases," the a priori predicate (to chase) defines the action but does not limit the subject's freedom and creativity: the cat may chase birds, balls, or butterflies. If the potential for action is internal and its predicative directions predefined, it is yet neither self-referenced nor self-contained: when the potential translates into action, it may produce multiple outcomes. While the human condition derives from its phylogenetic legacy, it does not result from phylogenetic incompleteness, but from phylogenetic redundancy. It is our exuberant tendency that nurtures our creativity. Admitting hybridization forces us to reject the idea of *téchne* as compensation and empowerment. The latter would imply the existence of a human essence that is intangible. It might as well be clothed, empowered, or extended, but it cannot be changed.

In the humanistic reading, the human can exclusively be Promethean, hence somatically incomplete, deficient, larval, nonspecialized, and ancestral (Bolk 1926). By attributing human superiority to its physical deficiency and assigning to its much-celebrated ontopoietic dimension the miserable role of a trickster (Hyde 2010), the Promethean reading offers a spectacle of ontological acrobatics. The human resembles an envious swindler who needs to devise new tricks to invent an adaptive space for himself. For the humanists, Promethean virtues are like the talents of an introverted mind, who needs to find inside itself solutions to pursue its goals. This interpretation completely excludes the *Epimethean contribution* to human inventiveness. For the humanists, technopoiesis, our inventive capacity, is solely the fruit of the human imagination. Imagination is totally self-referential: it is a journey of the mind in a sort of hyperuranion disconnected from the earthly dimension. It is a shipwreck of the human into the human, a departure from the *convivium*, which relieves creativity from any debt it may have contracted with the outside world.

Consequently, the technosphere becomes like another world, of which the human considers itself the demiurge (White Jr 1967). This is a misleading view that nurtures arrogance. Indeed, like a *fil rouge*, a sense of snooty superiority runs through the whole course of humanism. It was not so dangerous as long as the human technological potential remained limited and scarcely operational. But in the course of the twentieth century things radically changed. To consider imagination as an event without an external basis or support, as not inspired by the outer world, leads to viewing creativity as the most authentic expression of human isolation: alienation as an illusion of existential autonomy (Taylor 1991). For posthumanism instead creativity is a dialogical moment that stimulates us to renew our Self through hybridization. To account for technopoiesis as a moment of renewal and reconfiguration of human geometry through the highest expression of *thaumazein*—the epiphany or inspiration drawn from the world—and as a shift of status, which is not achieved but is in the making, means to change the foundations of our idea of *téchne*. Epiphany is the most accomplished expression of dialogical creativity and hybridization (Marchesini 2022). It is an event that engages the subject's interiority and emotional character by sensitizing it and giving it an ontopoietic thrill.

Imagination, therefore, is neither a retreat from the world nor a confinement within the human boundaries. It is the reworking of encounters with otherness made possible by copulative actions. The humanist view of creativity that considers imagination a withdrawal from the world, on the other hand, does not only encourage autarky. It actually causes emotional detachment from nature and the world and confers to *téchne* an eschatological meaning (Noble 1997) that is the opposite of *convivium*. What gets lost along the way is the copulative principle, which, in its *need-for* and *self-giving-to* otherness, confers maximum existential value to the fragility of the individual (Spengler 1934). Even hedonism pursued for the sake of itself, and life conceived exclusively as a search for pleasure, will lead to dissatisfaction. Hedonistic intoxication might well grant temporary satisfaction (Pulcini 2009). However, it will never suffice to make a life meaningful, if it is based on individualism and the illusion of autonomy and lacks external points of reference. To dispute the concept of creative autarky, of imaginative solipsism, is propaedeutic to the development of a new sentimental education. This type of education will reconnect the human and the world and recall the debts we have contracted with it for our cultural inspiration. Sentimental education also offers a new way of thinking about *téchne,* which turns out to be both the fruit of relationships and the *medium* of new ones.

## TWENTY-FOURTH THESIS

For posthumanism, technopoiesis is never compensatory but, on the contrary, it produces a dependence, which must be considered a consequence of the copulative experience, an a posteriori that defines it as a bond and as the need to reach a new existential dimension.

Overcoming the idea of human incompleteness means shifting from the principle of an original deficiency as the cause of technopoiesis and cultural production, to a *perception of deficiency* as a result of technical and cultural experience. This is one of the most important differences between humanism and posthumanism as far as the anthropological question is concerned. This paradigmatic shift explains a profound change in how posthumanism regards human nature, but also the contract of technomediation. Technical experience, and cultural experience in general, does not lead to emancipation. It creates a profound bond with the technical and technological referent. It is a perspective and performative connection that increases human needs (Hughes 2006). We feel this dependence, but we make an error of attribution: it is not deficiency that triggers technopoiesis, but we feel deficient as a consequence of technomediation. In other words, we exchange an ex post condition, namely a consequence, for an a priori need, namely a cause—a blunder that can be attributed to a rather common explanatory bias.

The error stems from the fact that our life dimension was created with the support of cultural and technological tools. The mere thought of being deprived of them makes us feel inadequate and naked. Yet this is an a posteriori feeling originating from the technomediated experience. Hence, we can speak of a *techno-addiction*, namely of a real dependence, which started when the use of tools changed our points of reference in life. Any cultural or technical apparatus modifies our perspective and performativity (Arthur 2009), with considerable repercussions on how we perceive the effectiveness and efficiency of performance itself. The hybridizing act is a process of inclusion—more correctly, of *existential metamorphosis*. The hybrid qualities acquired through an apparatus that the subject has made its own are attributed to the subject itself. Whenever the human invents a new device, the latter is introjected, that is to say, the human somatizes the qualities of the device. Hence, simply thinking of being without the device causes a feeling of lack. Posthumanism, therefore, flips the explanatory model: technopoiesis does not compensate for a deficiency, but produces it. Rather than an actual deficiency, it is a feeling of deficiency: the device does not provoke biological incompleteness, but creates a need (Balibar 1996) that is experienced as a deficiency. Twentieth-century generations know very well that they did not use to miss the smartphone. It was the advent of this tool that actually created the dependence from which we suffer today. It is a type of dependence that risks misleading us, especially with regard to supports that were introjected in the remote past, because we have no memory of how we would do without them. It is a huge mistake to conceive of technomediation as a compensatory and emancipatory tool. Technomediated events create addictions that, as we shall see, are much more articulated than mere performance.

The humanist explanation is naive, but it does support the idea of human purity, albeit immersed in a vast technological framework. If we consider human destiny as a progressive emancipation from the telluric, the external apparatus also plays a purgative role. The function is externalized, not performed directly by the human, who remains unspecialized (non-conjugation), preserving its fundamental difference from the brutes (Portmann 1936). Entrusting the "dirty work" on the world to an external support helps perpetuate those neotenic and larval features that are at the core of Botticelli's aesthetics: infantilism, gentle hatching, levity, and remoteness from the shaggy and bestial animality, all traits in line with the essential human core. As postulated by Pico, human dignity consists of indeterminacy. We may observe that, in order to purify the human from the residues of animality, still visible in its somatic dimension, humanism adopted an alchemical procedure that betrays the influence of hermeticism. Technopoiesis does not only need to compensate a deficiency, but also to erode all residual bestial traits. Such a blunder is even more apparent in transhumanism, which views technomediation not only as human empowerment, but also as a process of transcendence (Kurzweil 2005). Abandoning the organic is often described as a process that converts the human into pure information. The human becomes more suitable for being transformed, endowed with the fantasmatic potential of simultaneously inhabiting multiple bodies and less subjected to somatic constraints.

An incorrect reading of human nature and the ontopoietic process, which in transhumanism is not even anthropopoietic, can only lead to fallacious arguments, such as that of human emancipation. We brag that technoscience will liberate us from today's constraints, without even realizing that every hybridization causes dependence and, somehow, makes us more fragile. Yet posthumanism views human fragility positively, because in fragility and dependence the human expresses its best qualities (Levi-Montalcini 1987). The humanist tradition that is based on human existential self-foundation instead pursues autonomy and nurtures arrogance and dangerous behaviors, encouraging our worst side. It also tends to view technology as its armed wing, which helps the humans pursue their narcissistic and greedy goals (Von Goethe 1912). These traits do pertain to human nature. However, today they are emphasized by a sentimental miseducation. Posthumanism instead considers fragility the best antidote to arrogance and solipsism (Caffo 2022): it kindles convivial and compassionate propensities, nurturing the existential dimension of *Being-with*. Knowing that technopoiesis inaugurates a horizontal movement of dependence corrects the deficient view of human nature. It also renews entirely our view of technomediation. Instead of seeing the technopoietic event as a tool to compensate for a deficiency, we should refer it back to desire and our copulative exuberance, namely *our tendency to go beyond ourselves and ask otherness for help.*

Viewing human nature as deficient is inextricably linked to a Promethean and autarkic conception of technopoiesis. If we consider the technopoietic event an expression of copulative exuberance instead, the principle of autarky crumbles. Creativity originates when the human opens up to the world through a *rite of contamination*. It should be considered a rite of passage (Van Gennep 2019) officiated by otherness, whereby the human reaches a new existential dimension. Dialogical creativity transforms the relationship into a predicate-constitutive phenomenon (hence *constitutive relationship*). Otherness plays a central role in the process whereby new *téchne* or a particular cultural habit is introduced into the human dimension. Our relationship with the outside world is a great motor of fantasy and imagination: having a striking, disorienting, and decentralizing experience makes us dream of different and possible existential dimensions. This is not a contemplative experience. It is an engaging and overwhelming, gripping and revelatory event. It is an *epiphany* (Marchesini 2018), an event that transcends its condition as a phenomenon and takes hold of the subject, dragging it into a completely different dimension. Whoever experiences an epiphany encounters something *other-than-himself/herself*, something that literally invades him or her and unleashes a different perspective. The poietic event is always an epiphanic phenomenon, a sudden illumination. By encountering a perspective-performative otherness, such as, for example, a flying bird, the subject identifies with it and realizes that a new existential dimension is possible. The epiphanic process is not mere imitation, but an introjection of otherness and a projection into otherness. In order to understand epiphany, we need to return to the hybrid as an ontological dimension. Hybridization does not add new performing potential to what the entity already has. Hybridization is a metamorphosis of the entity, whose results are unpredictable and uncontrollable.

We are also mistaken if we believe in the fixed directionality of technopoiesis, whatever its epiphanic origin. Every single hybrid emergence, in both biological and cultural events, always lends itself to *ex-aptation* (Gould and Vrba 1982), to unpredictable functional co-optations. By means of example, let us think about the evolution of information technology and how we use it. The history of the computer resembles that of bird feathers—from thermoregulation structures to flying tools! The posthumanist climate might be imputed to the achievement of a certain technological threshold or the advent of a singularity (Vinge 1993), but I disagree. Posthumanism was born when, at the end of the nineteenth century, next to the unquestionable historical merits of humanism, we also began to see its limits. In other words, posthumanism began with the emergence of a new interpretative framework of the human. The cultural climate that developed in the "short century" heralded posthumanism, because it demanded new answers, which modernity was unable to give. Darwinism as a laboratory of thought that went beyond Darwin himself created the basis for mending the rift between the human and nature, from which humanist hermeneutics derived. The advent and establishment of eco-ontological thought, based on constitutive relationality and transindividuality (Balibar and Morfino 2014), has eroded some humanist assumptions, such as the idea of a self-referential and self-founded subject. The reinstatement of a historical and indeterministic dimension of evolutionary processes, and the birth of complexity theory (Von Bertalanffy 2015) were important additional elements in this paradigmatic metamorphosis. In philosophy, Deleuze enhanced the importance of difference as it had emerged from the vitalism of Nietzsche and Bergson (Bergson 1911). Admittedly, to think of posthumanism by only referring to technological excess provides not a trivializing, but a misleading picture. Yet it is the very issue of technology that highlights the incongruity of the humanistic explanation because it relates technology to human deficiency, solipsistic creativity, and performative outsourcing by means of ergonomic tools.

Let us return to hybridization as the pivotal difference with the humanist paradigm. A tribal dance inspired by the courtship ritual of the grouse, or martial arts poses that draw on wolves' and bears' postures and kinetics are not the plain repetition of choreographies or of the dynamism observed in otherness—the animal that is "good to think with" (Lèvi-Strauss 1971). They are, in fact, creative emergences. And they are not the fruit of an introverted imagination alien to the world, either. Admittedly, epiphany is typically human, because it is an expression of its copulative propensity. Yet it is not directly a human legacy, but the outcome of a striking encounter. Consider the following example. If creativity is a moment of inspiration and revelation, we must admit: (i) the existence of a desiring drive in the human; this makes the phenomenon so interesting that it becomes epiphanic, namely revelatory: it concerns us; and (ii) the active and co-factorial participation of otherness; since otherness is not completely alien to us, it resonates in us and can modify our existential geometry, as in the physical phenomenon of entanglement in quantum mechanics. Mimetic exuberance, our strong tendency to identify with others, creates the conditions for hybridization. Yet epiphany would not happen without the encounter with an engaging otherness. If creativity were not dialogical, the hybrid principle at the heart of posthumanist thought would not exist.

Let us now dwell on that perception of deficiency that for humanist anthropology triggers the poietic process. Every hybridization is, in fact, a decentralization (Onishi 2010), because it moves away from a phylogenetic center of gravity. It also creates a dependence, because we have to rely on otherness to achieve a particular existential dimension: after the hybridizing event, the newly acquired existential dimension cannot be reached autonomously. Let us return to flight as an existential dimension inspired by birds (Sax 2021). An ornithological epiphany allows the human to imagine what hovering in the air feels like. Only after this imaginative experience do humans have the urge to live and experience this particular existential dimension. This urge is always a consequence of epiphany, an experiential a posteriori. And the perception of deficiency is always the consequence, not the cause, of hybridization. Posthumanist anthropology needs to avoid the circularity created by taking the consequence of the copulative experience—the perception of deficiency—as the cause of the experiential event. To do this, we must admit: (i) that intrinsic human exuberance leads to openness, (ii) that exuberance causes strong copulative engagement, and (iii) that the subject's encounter with otherness, whatever this may be, produces hybridization and consequently decentralization and dependence. The principle that explains technopoiesis is thus reversed: from deficiency as a cause, to the *perception of deficiency* as a consequence, the cause being exuberant and redundant human nature. Expressive exuberance is copulative, because human nature is fascinated and attracted by the world. This fascination derives from a sense of correspondence that makes the human feel in tune with other living beings and see itself in them.

If we manage to go beyond the compensatory explanation, we understand why technopoiesis produces dependences: the more humans have evolved culturally and technically over time, the less autonomous from the world they have been. It is not the other way around. As poiesis produces a centrifugal thrust and causes decentralization, the human system becomes increasingly fibrillating and enters a condition of nonequilibrium: decentralization means reaching increasingly unstable positions (Nye 1996). Rejecting the compensatory and exonerative explanation also clarifies this process' exponential rather than decelerating progression. If technopoiesis were the response to a compensatory need, our feeling of deficiency should progressively decrease. Instead, if it produces dependence and increases our needs and aspirations, the system's accelerating trend and our deficiency feeling are justified. The compensatory reading (Gehlen 1988) makes us think that culture initiates an anthropo-centripetal process and helps stabilize the human dimension. If we assume that culture is the fruit of an exuberant condition instead, the trend is anthropo-centrifugal, which is what we can actually observe when we consider the extent of human technical evolution throughout the modern age.

Let us summarize. The anthropopoietic event is not triggered by deficiency, but by human copulative exuberance. The human dreams through otherness. The resulting experience is always dialogical, the result of the convergence of human predicates and the world's distinctive traits. But this process does not merely produce additional traits. Epiphany reworks perspectives innovatively and productively. Our symbolic abilities can pry apart different references and reshuffle and reinterpret them out of their specific perimeter. Flight is no longer just a performance that displaces the body but an existential dimension (Marchesini 2017). It may stand for lightness, ascent, gracefulness, and spirituality, transform into dance or angelic features, become a metaphor for will or thought. To take on wings, therefore, means to acquire new, previously inaccessible, ways of being. The epiphanic experience is productive because it gives access to new continents of meaning. The epiphanic experience also creates a semantic and existential bond between us and the referent: for example, it will no longer be possible to access the aerial dimension without the intermediation of birds. Technopoiesis is, therefore, a creative process originating from a copulative need: our dreaming by means of everything that surrounds us, our introjective imagination, which continuously enriches itself and quenches its thirst at the source of the world. The phenomenon becomes epiphanic when otherness transforms from an *other-than-oneself* into an *other-in-oneself*. Epiphany is creative because, once experienced, it gains a dimension of its own in the human imagination. It is no longer a physical experience, but transforms a large number of figurations and compositions that are all mediated by otherness as a tutelary deity. It is a process that can find multiple expressions—let us consider dance, for example. Once a particular choreography has been observed in a particular species, it is never reproposed exactly as such but transformed and adapted to human characteristics. Thus, it gains ever-richer symbolic values and prompts further hybrid and generative processes. In some dance choreographies, for example, dancers imitate the movement of the head of hens and cockerels. Combined with other movements, dance choreographies can be compared to bricolage or patchwork. Assembling together bits and pieces creates an entirely new and original piece of work, a choreography that acquires a life of its own.

With regard to dependence, we need to consider several things. The easiest point to understand relates to performative standards, namely the perception of optimality brought by technomediation: if we have gotten accustomed to using the mobile phone, using the mobile phone becomes our standard, something we cannot do without. Technomediation not only improves functions we already have—sometimes by enhancing them, other times by extending them—but it also creates forms of access and performativity that do not simply improve an already existing function, but are totally innovative. The human gets accustomed to these functions. The technomediated performance, therefore, sets an acceptability level of the function, which exceeds what can be attained by our body alone—hence the perception of deficiency. In addition, every piece of technology is an entity connected to a system of technomediation. Accordingly, there is no technology that does not require a technique, or is disconnected from other technologies. A person does not wear a piece of equipment, but lives in a technomediated context (Carr 2010). Therefore, an instrument cannot just be considered a disjointed entity responsible for a specific change in performative standards. Every single invention and innovation changes our entire technospherical niche through processes of connection and functional emergence that vary from area to area. Communication and computation technologies, for example, used to be separate, but have now merged. We can no longer refer to the telephone today by using the categories that were applied until the 1970s. But this is also impossible for the computer as we cannot separate the two devices. This is why a piece of technology cannot just be removed. It must be replaced by an update or an invention that may enlarge the performative domain of its original function.

Rather than simply enhancing or extending human performance, technological revolutions change people's living dimensions (Anders 2002). The systemic dimension of the technosphere creates a niche that goes beyond single performances and inaugurates a lifestyle where different media tend to harmonize comprehensively. Another example comes from the digital revolution, which has radically changed how we use many services that used to be provided by analog devices. The analog experience differs totally from the digital one. It does not change the content of a music compilation, a film, a photograph or a book, but it demands an immersion that separates the subject from the medium. Making a photograph with the analog method meant touching the film, smelling the acids of the darkroom, working manually, and even using particular techniques. All this gets lost in the digital, which admittedly offers new possibilities. The number of photographs that can be stored in the *cloud* or in the memory of our computer today is more than we can ever look at. Children born in this century have hours and hours of video recording almost every single day of their lives. The memory of their past acquires a synchronicity that was unimaginable even just twenty years ago. So, is it possible to address the technosphere only in terms of performance? Obviously not. Technologies not only change our experience, they actually define the modes in which we participate in the social community too, because they define participatory standards. Can young people do without the smartphone today? No, they cannot. Not because of the specific performances the device offers, but because it sets out the rules of the relational *convivium* and, sometimes, even affects social acceptance. We should not focus on performative aspects, therefore, because technology often works like a pass: it allows to be part of a group and share a generational dimension. Indeed, different age-groups are characterized by the use of different technologies; they are interested in ways of communicating that differ from each other and, accordingly, produce in them different experiences (Niessen 2013). For example, the generation born before the year 2000 prefers Facebook as a social media platform, while the millennium generation uses TikTok. The former communicates primarily through posts, the latter through videos. The experiences of the different generations differ not only in content, but also in aesthetics and rhythm. In cinema, for example, today's devices and techniques of direction and production change perspectives in the time sequence and in the narrative rhythm, inducing experiences that can hardly be compared to those of the past (Demichelis 2019). Meaning is not only derived from the content of an experience, but also from its rhythm and duration, namely from the speed in the sequence of events. For example, when I travel by car, I experience the landscape differently from when I cycle. Finally, we should not forget that all technology also fulfills another task: it represents different facets of an individual, his/her conformity as well as his/her singularity, so that we should actually speak of identity apparatuses.

## TWENTY-FIFTH THESIS

The concept of technique and technology changes profoundly in posthumanism, because it is no longer seen as an external empowering tool, but as an entity that inhabits the body, transforming its possibilities and projections.

Experiences made in our youth create a sentimental memory and education that profoundly affect our identity. The cultural and technospherical niche in which we grow and create our existential references defines our experiential milieu. This is what we have called proxemic identity. It defines not only the performative, perspective, and participatory inclusions, realized through technomediation, but also the conjugative structure, the type of bonds on which we base relationships. For example, social media create friendships and connections that cancel the spatial dimension. In the future—perhaps not even a too distant future—the presence of a real interlocutor might no longer be needed: an avatar on the net, endowed with biographical content, will stand in for someone who died (Sisto 2020). Although no longer biologically alive, he/she will continue to exist on the web, posting and interacting with followers.

When we speak of a sentimental education, however, we refer to the bonds established in the developmental age. Just as the clod where a tree is rooted is part of its organism, neither more nor less than its roots, trunk, and crown, these bonds define the overall proxemic bubble of an individual's feelings and desires. Our experience can be spatially represented as an extension of our identity. But it would be more correct to represent it as a network of capillaries that extends beyond the skin, fueling an extracorporeal circulation of vital importance for a subject's emotional identity. Today's children, who grow up immersed in digital fluid, have a different proxemic extension than the so-called baby boomers (Campo 2020). The latter spent their childhood in an analog kaleidoscope of manifold objects, each providing specific services and requiring equally specific techniques. Contained in a world overflowing with functional objects, baby boomers also collected experiences that had no resemblance to those of the generations preceding them. Still immersed in a rural context of plants and animals with different rhythms and other smells, these generations had a much closer contact with the ground. Even their hands showed different praxic conjugations (i.e., specific manual competences) and epidermal morphologies. Proxemics is made of memories, and memories are not simply traces of the past, heaps of sedimented information. They are actually distinctive emotional styles.

Sentimental education is the expression of the copulative contract stipulated in the developmental age, through the choice of specific representative references of the Self. These references define what is welcome and familiar to the subject. The subject is ensconced in them, like in the maternal uterus. These references conjugate the subject's verbal predicates because they are complements of prospective actions (Bove 2002), they are introjected by the subject as part of the Self. And, in a way, they are part and parcel of the Self. However, if we read experience merely as content, as if our memories were a home we inhabit, we develop the wrong idea of experience. This anastomotic structure profoundly affects feeling: all experiences produce not only a cognition of the world, but also a metacognition of the Self (Freud 2004). Devices are interlocutors and interfaces. We do not only get accustomed to using them—we get to know their characteristics, experiment with their functions, and develop the right techniques and practices—but we also get to see the world through their focal lens. Reality is deformed by technological devices, so the experience we make with otherness also changes, because it is our whole perspective and epistemological panorama that changes (Fukuyama 2002). The device imposes a particular training onto our psychological components, which are subject to an evolutionary differential. Arguably any device, especially when it affects an individual during the developmental age, has a formative, rather than just an informative, impact.

This view introduces an interpretative leap that is also supported by scientific evidence. Technology modifies the functional structure of the body, so that it is correct to speak of a *techno-physiology* affecting our entire somatic structures (Marchesini 2017). One of the most evident changes caused by today's multimedia technology is our posture. We tend to lean the neck, and curve the back, forward. Our entire somatic *chassis* is reshaped, with reduction in respiratory amplitude, muscle hypotrophy in the entire posterior train and compression of the vertebrae. We can observe again today what used to happen in the past, when physical jobs that were particularly straining permanently altered the body's biomechanical characteristics. Although we may be unaware, every technopoiesis is also an anthropotechnique, because our relationship with the device always reshapes our body. Let us think of the muscle-tendon structure of our feet. Overused after the Industrial Revolution, today's sedentary life and prolonged standing in front of machines has weakened it with a knock-on effect on other parts of the body such as the knees, pelvis, and cervical structure, which become deformed (Cregan-Reid 2018).

Computer work is not risk free, either: we know that the endocrine setup is modified mainly through the activation of the sympathetic system, with considerable repercussions on stress mechanisms. Even our attention system changes: the reactive component increases by comparison to our ability to concentrate. Because of the overstimulation of the retinal-epiphyseal system, the wake-sleep rhythm changes, modifying the production of melatonin. All neuromodulation factors are reset, especially the dopaminergic and serotonergic ones. Because of today's dominant informatic reductionism, we are used to thinking about brain functionality only in terms of synaptic wiring. We do not realize that the mind emerges from the body's overall metabolism—with the endocrine system playing a relevant part. Let us consider, for example, the mechanisms of reward and expectation triggered by dopamine and elicited by the use of social media (Greenfield 2015), or of the relationship between aggression and vasopressin stimulated by some electronic games. Yet technology-induced renunciations also make an impact. Let us think of the feeling of security produced by oxytocin, which is released by convivial behavior, and of the role of endorphins related to physical exercise. The type of physicality introduced by digital technology revolutionizes all this.

We are only just beginning to assess the impact that the lifestyle introduced by computer technology is having on human physiology. When we speak of an ontoplastic effect of *téchne* we refer to the fact that any device trains different parts of the body, as clearly happens in the muscle-tendon system: excessive effort may compromise its structural and functional integrity as much as underusage or a reduction in its use occurs when a device replaces an organ's previous function (Spitzer 2012). The use of the calculator has undoubtedly thwarted our ability to do mental arithmetic, as much as our memory has collapsed with the advent of mobile phones. There are situations in which the apparatus affects the functional organization of the organ. We can observe it in the way we use our hands. Since an original cubital-palm grip, the hand has developed new functions: oculo-manual coordination thanks to technical work, flexibility thanks to craftsmanship, the clamp grip from the age of school. Today's use of the smartphone has also led our hands to develop a two-thumb synchronous functionality.

Rather than of an extension-amputation of an organ (McLuhan 1964), we must speak of a functional shift that defines new and different terms of reference for the body's coordination and organization. When the device separates the synchronous functionality of two apparatuses, possibly even creating new complementarities, we may speak of a dissective effect. Unlike what happened in the past, for example, the way in which we visually perceive movement today is disconnected from the locomotor system: it is connected to praxias because our manual movements refer to our movement in space, as exemplified by how we use our arms to move in virtual reality. There is a change in the prevalence of our senses. This is obviously not due to transducer changes—at least not to a large extent. It is due to the excessive use of sight and hearing and a parallel reduction in smell and touch: the cognitive density in the brain areas where these sensors are projected changes. This training defines the interface qualities of our organism, which we then carry along in all our life experiences. The immune system also plays a crucial role because, through inflammatory mechanisms and the release of cytokines and histamine, it has repercussions for our psychic condition (Sapolsky 1994). The state of mental tiredness associated with computer work might be linked to these aspects.

By modifying the mother's endocrine and metabolic setup, the technosphere is a niche that may influence not only the developmental age, but the prenatal period as well. It is known, for example, that heated plastic can release substances that affect the hormonal mechanisms involved in defining the embryo's sexual characteristics (Colborn et al. 1996). Even though research on the influence of the niche on embryofetal development, the so-called evo-devo effect, is still at a very early stage, important evidence shows that, through multiple epigenetic factors, the living environment affects genetic translation. It is sufficient to silence or weaken a gene, delay or anticipate its transcription or, on the contrary, enhance its expression, to modify the phenotype and its performative and metabolic incidence. We know that, in other species, the different meteorological, physical, and chemical conditions of the environment produce morphologies that are in many respects divergent, such as sexuality, the color of coat, the adipose tissue, the shape of the skull, and thermotolerance. In humans, environmental factors cause predisposition to cardiovascular problems, diabetes, and numerous neoplasms (Barker 1997). This infiltrating view of technology changes our way of considering it.

In the humanist reading, *téchne* is like a garment, at the most it has the function of a probiotic: it may facilitate certain functions and enlarge the range of preexisting predicates without affecting the human essence. Technology can be put on and taken off like an armor that preserves human integrity or a glove that protects the hand to avoid contamination. These images are still present and vivid in transhumanism. This is why it is easier for humanists to understand the transhumanist approach than the posthumanist one. Posthumanism instead considers human predicates the outcome of the decentralizing processes produced by technopoiesis (Pepperell 2005). While humanists believe that technology frees and purifies the human by externalizing its conjugations, posthumanism subverts the whole human system, because *téchne* has an ontoplastic effect. The latter aims at changing: (i) the human affective, homeostatic, aesthetic and dispositional setup; (ii) the physiological organization of the body, in its endocrine, immune and metabolic systems; (iii) the cognitive structure and, more broadly, the type of intelligence; (iv) the architectural, somatic and performative configuration in terms of both structure and performance; (v) the subject's perceptive and epistemological interface; and (vi) human goals. In one word, our entire existential dimension. *Tèchne* is not a garment. But it is not an amplifier, either: it modifies human predicates as a whole, so much so that what follows the advent of a form of technomediation is the coming into being of a new man (Arthur 2011). *Téchne* is like a virus: once it has entered a cell, it reprograms and flexes it, enabling new ontoplastic translations. This is the reason why posthumanist philosophy questions the ergonomic view of the technical equipment: it is the flesh that always adapts to the device.

The type of synaptic wiring *tèchne* generates offers a very appropriate example of its ontoplastic effect. As we know, both synaptogenesis and its opposite, apoptosis, are affected by the synchronous and consequential activation of various neurons (Edelman 1987). They build assemblies, or connection networks, responsible for the subject's psychic states. The nervous system is a dynamic entity that produces networks with a high combination probability within the connectome. They have different degrees of stability—a metastable system again—and change with the subject's experiences. Spending prolonged time in front of a particular technological interface, such as the digital one, affects our synaptic wiring. No wonder the cognitive characteristics of today's generations differ from those of previous generations. Some functions have been radically depowered, such as memory, calculation skills, concentrated attention, reflective ability, and communicative complexity. Others have been boosted, such as responsiveness, multitasking, widespread attention, the tendency to make quick connections, and the ability to navigate the infosphere (Lévi 1990). This is the reason why we must stress that there is a remarkable difference between the transhumanist attitude, which is overall techno-enthusiastic, and the critical stance adopted by posthumanism. Critical means problematic and precautionary. If it is incorrect and anachronistic to encourage forms of neo-Luddism, it is equally foolish not to see that, alongside the benefits of technology, other aspects should also be considered. Technology changes us much more than we are willing to admit. Our anthropocentric presumption makes us believe that we have control over the machines and we can preserve the purity of our center of gravity. We think that the function of technology can be downgraded to that of a tool or gadget, something that remains outside of us, confined out of our body, like barbarians were confined outside the *limes* of the empire. This is not the case. Technology is always infiltrating; it is a mutagenic factor. And induced mutation is never completely predictable and never ceases to surprise. Maintaining a critical attitude, therefore, is not a title of merit, but a sign of wisdom.

## TWENTY-SIXTH THESIS

Transhumanism has an irenic and sometimes salvific view of technology, considering it to be a human expression and thus under human control; posthumanism, on the other hand, has a critical view of it, regarding it as much more unpredictable.

To have a critical view of *téchne* means to have a critical view of its technopoietic, technical, and technological expressions and, in general, of its systemic structure, the technosphere, which configures itself in the human relationships with this group of devices. To be critical, however, does not mean to be against the production and use of machines and devices. It simply means to assess their impact in global terms. In the frenzy for progress that often turns into a kind of foolish intoxication, many problematic aspects (Spengler 2021) require serious consideration. To take a critical view does not mean to censor, but to ponder. It does not mean to adopt a technophobic or neo-Luddite attitude, but to try to examine as much as possible what the social consequences of an invention-innovation are in order to find solutions. The enthusiastic and celebratory approach of many transhumanist authors is not far from the humanist motto that makes of every human product a celebration of human dignity. What it comes to in the end is that whatever is technically possible is also legitimate. As previously observed, posthumanism does not mean overcoming the human species or entity, but defining new points of reference in order to go beyond anthropocentrism.

Therefore, I would like to underline some of the problematic aspects that are linked to the impact of new technologies: (i) differences in access to them increase social differences and contribute to phenomena of marginalization and exploitation (*technological divide*); (ii) a phase or a period of mental numbness that can result in hallucinations or loss of self-control often follows people's initial lack of ability to handle a new piece of technology (*technological ecstasy*); (iii) the device is used by the establishment to exercise control: people are made to believe that they share technology while their attention is actually diverted from noticing the exercise of power (*technological subordination*); and (iv) the subject becomes dependent on new technology and feels compelled to comply with the procedures and living standards it imposes (*technological slavery*). Individuals may think they have more time available, but this freedom is thwarted by social constraints (e.g., bureaucratic constraints such as the need to memorize passwords) as well as by the need to conform to the rules imposed by technology: (i) new cognitive and performative abilities are required, which widen generation gaps, causing social breaches and misunderstandings between parents and children; (ii) additional dependencies from other technological families are created increasing our dependence on the technosphere, as highlighted with reference to the relationship between telephony and informatics; (iii) new constraints are imposed on the body with resulting dystrophies, metaplasias, and neoplasms caused by environmental alterations; and (iv) changes in interhuman relationships and their repercussions on the body cause ethical problems, not to mention how every new technology modifies the characteristics of warfare in conflicts.

These aspects should not be ignored, denied, or underestimated, but they should not engender diffidence or fear, either. Taking a critical attitude toward both human nature and technological progress basically means considering both their risks and benefits. It means avoiding emphasizing the former by adopting an attitude of total precaution, because not progressing also comes at a price. Accordingly, I would avoid adopting precaution as a compass (Jonas

1985). At the same time, however, we cannot indulge in enthusiasm too easily: we should not only look at the benefits that every technology inevitably flaunts, and depend on the neophilia that is typically human and emerges at every opportunity for progress. Technoscience cannot be a faith, because it hinges on an epistemological principle that is implicitly metastable: this is demonstrated by the asymmetry between falsification and verifiability (Popper 2002). Transhumanist hyper-Enlightenment ramblings about how reason, science, and technology will usher in a new age of Enlightenment and free the human from the last remnants of its biological constitution are a real problem. The transhumanists' proposal for the future of humanity has plundered science-fiction narratives. Moreover, transhumanists have followed uncritically but, most of all, prejudicially, the reductionistic, explicative, and descriptive readings of physiology. For example, they have embraced the cognitivist conception of the mind as an information package that runs on a hardware made of flesh. This double error—science fiction taken for reality rather than for transreality, and reductionism for a criterion and explanatory model of the biological phenomenon—is a most serious distortive risk. It is also a very dangerous illusion and temptation, because it is most palatable for a human nature and cultural dimension that, like the humanist one, aims at exercising control. The proposals of these authors generally focus on the specifics of their imagined reality, such as immortality or the transition toward a postorganic dimension. Therefore, what is most important, namely philosophy, fades into the background. In my opinion, the most serious problem is the lack of appropriate philosophical reflection on the transhumanization of the project (Huxley 1957). It is based on the illusion of full human control over nature and on a transcendent idea of the human. The posthuman condition is often incorrectly described and discussed by transhumanists because they claim they can simultaneously anticipate, hence control, the future and transcend it. The question that arises is: how can I imagine what my goals are in a dimension that I can no longer call mine because it transcends me? In other words, if I think of a human, who has been disembodied and become pure information, how can I possibly imagine its desires? While posthumanists believe that the human has always been posthuman, that the human has always been a hybrid, for transhumanists, the posthuman condition is yet to come. Therein lies the paradox: transhumanists describe a posthuman condition that preserves wishes, aspirations, and feelings that are human while conceiving an entity that is totally different from the human. I cannot possibly expect such an entity to have aspirations that are the same as mine. Therefore, I cannot understand how a similar philosophical approach can be taken seriously: not because of its marvelous prophecies, but because of its basic and blatant structural inconsistencies.

We speak of an affective condition or, better, we assume the existence of an affective, desiring or sentient drive without realizing that these characteristics pertain to the diachronic condition of the organic. This means that we cannot speak about, or explain a biological phenomenon by adopting a synchronic focal lens, nor think of time as a Newtonian constant. We cannot separate the *Being-there* from the diachronic condition implicit in every snapshot of the subject's life by referring to subjective expression in isochronous terms. These blatant errors demonstrate how shortsighted these scholars are. They totally ignore the condition of the living. I believe the reason for this error to be much more trivial, however, which makes it all the more necessary to expose it. To imagine a human who can be equipped with almost unlimited performative skills and even be maneuvered remotely, because it is totally immersed in the flow, as in the case of ambient intelligences, recalls and, I daresay, tickles the imagination of the war industry and attracts huge funding. This is the reason why there is little hesitation in rambling and betting on the infinite possibilities of a technoscience, that, so far, does not seem to have disappointed human expectations. In this case, science fiction becomes—though it would be more correct to say that it *is interpreted as*—the vanguard of technoscience rather than the expression of a narrative fantasy whose predictions may sometimes be anticipations, but also clamorous mistakes.

The transhumanist proposal is often considered to be the maximum expression of technoscience and the faith in it. This cannot be true, however, because the two terms do not overlap. Science is problematic (Popper 2021), and expressly refuses fideistic approaches. A scientific theory is considered valid only until it is disproved. It is correct to state that, in some circumstances, science fiction has been able to predict the future. In these instances, however, it failed to indicate correctly the means and consequences. Moreover, gross blunders were made when predicting comprehensive scenarios, which announced conditions or possibilities that proved to be blatantly wrong. A clear testimony is provided by the sci-fi movies of the 1960s. There are obvious recurrent themes, such as the golem, the *Doppelgänger*, the automaton, dystopia, or heterotopia, which have more affinities with magic than technoscience, and sometimes come close to religious or sacred experiences. To talk about science fiction as an anticipation of the future is wrong if we refer to the past, and it is blatantly out of place today. Contemporary science fiction does not anticipate technology, but runs after it, for a very simple reason: today's writers struggle to keep up with the speed of technoscientific advances. Posthumanism, therefore, does not avoid depicting future scenarios, but does so with caution, because of the countless factors that come into play. In the first half of the twentieth century, nobody could imagine today's reality, which has been revolutionized by the digital, and even a few decades later people were not fully able to grasp the scope of the transformation taking place around them.

The fundamental turning point of the posthumanist vision is its refutation of anthropocentric control over *téchne* (Kelly 1995), that is, the idea that *téchne* is unpredictable, introjective, and systemic. First of all, we have to acknowledge that every technopoiesis is never entirely in the hands of its designer, because of three main reasons: (i) firstly, because the technoscientific discovery is partly due to serendipity and its effect is epiphanic, so its scope is not strictly rational; (ii) secondly, because of the co-optative aspect of every technopoietic event, the fact that, similar to evolutionary *ex-aptation*, it inaugurates applicative possibilities that cannot be foreseen during its design (e.g., the computer); and (iii) thirdly, because of the emergence effect: while a technological device is conceived and designed for a particular purpose, once the techno-media area is created, it immediately hybridizes with other technologies and produces unexpected outcomes. These three basic considerations suggest that the effects and results of a technopoietic act cannot be entirely predicted.

Moreover, technology is always infiltrating, and it shapes the body through a reorganization of tissues, physiological metamorphosis and a process of somatic remodeling. It is not just a technical metamorphosis of the human even if all technology requires technical innovation of human performativity. What happens is a predicative shift that affects human ontology as a whole, including human goals and the existential purposes that become possible. It is impossible for a hypothetical posthuman entity to have the same goals and objectives as the contemporary human. This transfer did not work even for changes of minor proportions, such as the ones that occurred in the past century. Today we believe finitude to be a curse; the contrary might be true in the future: no one can know for sure. The introjective character of *téchne* shows that everything that was believed to be a mere external apparatus, serving human unchangeable goals and purposes, penetrates and modifies the human profoundly in all its facets. The crucial difference between humanism and transhumanism on the one hand, and posthumanism on the other, lies here: for posthumanism, the human is a mobile entity that can neither be entirely transformed nor conceived as static. Treating technopoiesis analytically, as the simple design and realization of instruments, means not having understood the systemic character of technomediation. It means adopting a fully analog approach, an approach that precedes even the digital revolution. Any technological transformation fits a certain milieu: it amplifies its systemic power on the one hand, and introjects its logic on the other. There can be no pure *extropy* (de Grey and Rae 2008), because resources are limited anyway, although it can certainly improve the productivity of processes and, therefore, reduce the overall entropy of the system. Posthumanism is the thought of complexity. For this reason, it urges us to look at technoscientific development systemically by adopting the logic of a system. This means not making marvelous projections into posthuman futures, but prudential hypotheses about possible scenarios that take into account, ecologically, all the factors at stake.

# *Chapter 8*

# Paradigmatic Evolutions

In the modern age, posthumanist philosophy made a paradigmatic leap forward. It started a profound metamorphosis in all major cultural fields, which has important repercussions and will require a very articulate discussion. The purpose of the four theses that complete this manifesto is to stimulate a debate about crucial issues, such as bioethics, aesthetics, scientific research, and epistemology. This chapter aims to provide the foundations for this discussion, defining a general approach and a framework, which, however, will require further work and a more specific articulation. Since it offers a new ontological model, the posthumanist transformation also has repercussions on values and orientations (Ferrando 2016), and paves the way for further research. This chapter summarizes more explicitly the meaning of this whole study, which underpins all its theses, being the need to avoid illusions and elicit a critical discussion about ongoing changes.

## TWENTY-SEVENTH THESIS

Posthumanism is characterized by a strong focus on the phenomenology of life and on bioethics, understood as a moral concern for the living world, whereby the human is given a duty of responsibility toward the biosphere.

Posthumanism has a profound interest in the phenomenology of life, as well as in our ethical responsibilities to it and what these responsibilities entail. The solipsistic drift produced by a vertical perspective, which condemns the modern human to suffer from loneliness (Benasayag and Schmit 2003), even when surrounded by people and in spite of all its technological devices, is one of the existential contradictions of modernity. The ideological anthropocentrism of humanism should not be confused with the anthropocentric perspective: while the latter derives from our tendency to anthropomorphize the real by projecting ourselves into phenomena, the former breaks our continuity with the world by considering technology a means to enact a second creation (Hughes 2005) and establish a human ontology founded on difference and exceptionality. Technoscience plays a central role here: it opposes both anthropomorphism and the anthropocentric perspective, as demonstrated by the way in which scientific counterintuitiveness (Wolpert 2000) can outdo so-called naive physics. Yet its purpose is to gain control over phenomena and it leads to reductionism in the interpretation of the world. On the other hand, ideological anthropocentrism, which features particularly in the modern age, can be defined as the cultural development of a self-centered perspective (Marchesini 2017). The latter celebrates the human as a special entity who stands in opposition to the biosphere and refuses to have anything to do with it. It is impossible to understand the contemporary age without taking into account the ambivalence described above: (i) on the one hand, since new technologies have challenged human intelligence, technoscience has cast a serious blow to the Vitruvian vision of the human as a measure of the world; and (ii) on the other hand, the new possibilities of accessing and controlling world phenomena, plus the fact that technologies can offer experiences that are an alternative to the phenomenology of nature, widen the gap between the human and the biosphere. The posthumanist vision is linked to the phenomenology of life (Henry 2008). For posthumanism *bios* is an entity that cannot be reduced to mathematical terms and translated into an isochronous process: it is always diachronic and creative. And humans are part of it. Thus, they cannot refer to the biosphere as if it were merely a cheap resource (Moore 2016).

A separative and autarkic human ontology ends up separating the human from the nonhuman on the basis of moral values: while the former is recognized as the purpose of moral action, the latter is only just the means. At the same time, human epistemological marginalization creates the contradictions today visible in transhumanism, which aims to transcend the human but draws on humanist ontological guidelines. With the advent of digital technologies, information in the form of the bit (Shannon 1998) has become the new primal element (*arché*) (Fredkin 1990) that overshadows the medium: if I conceive of the mind as a packet of information, it is almost irrelevant whether the information circulates in the brain or on a computer. Moreover, since in a quantum view any medium is made of matter, it can also be transformed into information. Digital technologies do not simply replace analog devices but actually produce important epistemological and ontological shifts. Our interaction with analog devices produced an experience that appeared to be merely ergonomic. With the advent of digitalization instead, every form of expression also produces a piece of documentation, a record (Bachimont 2018). Our relationship with the device is flipped: we can still say that we are going to watch a television program when we go online; but watching also means being watched (Ferraris 2020). It is correct to speak of a "philosophy of the digital epoch," because our interpretative points of reference of the real have changed: (i) from an epistemological perspective, in addition to the traditional deductive, inductive, and abductive processes, there is also simulation, which considerably reduces the need for human explanatory mediation; whenever I make a computer simulation, I can resolve problems and discover patterns without needing human intervention (Wolfram 2002); and (ii) from an ontological point of view, the idea of formation—definition of a form—through the split operated by the bit (i.e., the creation of discrete units) radically changes how we view the real: we perceive it as a sequence of separate elements. In other words, if we use the bit as an ontological principle, such as the 0–1 dichotomy, we no longer see the world as a *continuum* but as a discrete entity. This is what creates the ambiguous atmosphere that characterizes our times: while we have been ideologically manipulated in an anthropocentric direction like never before, the development of science has led us to interpret and understand the real and its phenomena in ways that are not solely based on human perception.

Posthumanism is the field in which these ambivalences are discussed. It makes us understand that, with technoscience, anthropomorphism loses epistemological consistency and anthropocentrism loses ontological consistency. In the eco-ontological view, the predicate originates from a relationship: somehow, eco-ontology brings together the ontology of the bit with that of the medium and views every expression of the real as the result of the relationship that exists and is implied at a given moment (Von Bayer 2004). This means that we cannot refer to the content of information in itself, or to a simple and dualistic *circulation of information* on a medium because information is always the relational expression that the real acquires at multiple levels: from energetic levels of quantum mechanics to the chemical ones of matter, from biological to psychological levels, from individual to social levels. Posthumanism is therefore a philosophy of relationships. There is no universal principle that makes the different forms of the real explicit. There are multiple levels that come into being through relationships. This multiplicity originates in the virtuality of the real, in its multifaceted potential (Kothari et al. 2021). For example, if I observe a piece of furniture and I am an antiquarian, what emerges from my relationship with it may be that it dates back to the eighteenth century. If, on the other hand, I am a botanist I might notice that it is made of a particular type of wood. The piece of wood has different qualities that are revealed by the type of relationship I establish with it. Ultimately, it is the type of relationship I have with the real that allows its qualities to emerge. This has important repercussions on our position in the world. Posthumanism considers anthropocentrism the final obstacle hindering human liberation. Freedom, however, does not mean breaking bonds in order to float in a solipsistic and self-referred dimension, but establishing relationships in order to be fully part of the *convivium*. This liberation must go through a number of stages that still stand in our way, starting from a new definition of the human, who should no longer be considered contralateral to the animal.

I believe that the human has enormous symbolic intelligence. However, I do not believe that this symbolic intelligence can be considered an ontological difference that justifies us viewing the human as something else than an animal. Nor do I believe that the symbolic intelligence singles out the human and makes it autarkic. The animal question originated as an ethical issue. Even today, when we speak about a critique of anthropocentrism, we imply that there is an ethical perspective. On the contrary, when I speak of a critique of anthropocentrism, I refer to all philosophical areas. This is why it is important to start from the origins of the animal question, which began with social movements, such as anti-vivisectionism (Ruesh 1975), but gained greater coherence with the advent of bioethics. Bioethics developed in the 1970s along with newly emerging ethical questions, such as: the environmental issue and the limits of development, the challenges to the notions of agent and moral patient, the whirling development of biotechnology and new ethological knowledge in the field of animal sentience and cognition. The same decade witnessed the emergence of a moral reflection upon interventions on life (e.g., genetic engineering, cloning, embryo splitting, and chimerization): the question that was raised was whether everything that was technically possible was also lawful. It was the dawn of a reform project that questioned consolidated assumptions. Ethical anthropocentrism, for example, created an aporia in the debate about the idea of the moral patient because it introduced a dangerous tautology: the right to own moral values was given to the human only, because it belonged to a special category rather than because it was justified by morally relevant categories (Regan 2004). Maintaining that the human is worthy of respect because it is human introduces a circularity into ethics that not only gives rise to prejudice—speciesism—but also contradicts the standard of reasonableness upon which ethics is based. Arguably, this incongruity exceeds even the arguments raised by anti-speciesism; it puts the foundations of ethical discourse in mortal danger. It is a bug that, if left to its own devices, could cause the whole building to collapse.

There is a close relationship between posthumanism and technoscience, although not in the humanist sense, based on the reductionist and mechanist explanation or on the use of technology as a means to achieve supremacy and domain over the biosphere. There is a close relationship between posthumanism and technoscience, because we need to start with what science and technology show us. Digital epistemology, for example, has shown that many of our difficulties in understanding the dynamics of complexity are due to our cognitive limitations. Other resolution systems make them anything but inexplicable (Ross 2001). The same can be said for ethology: it has revealed that there are different systems of orientation in the world. They use diverse sensory and cognitive systems and show how the texture of reality is much more complex and multifaceted than we used to assume (Hughes 1999). The emergence of ecology and its relationship with developmental biology have revealed the errors within the essentialist conception of the living, and have shown how the phenotype originates from multiple relational cofactors. Neuronal plasticity eradicates the distinction between body and technology by unraveling how the device becomes somatized and incorporated into the morpho-physiological matrix. In other words, thanks to new techno-scientific acquisitions, the foundations of the disjunctive and self-centered conception of the human are gradually collapsing. A relational view of entities questions the traditional divisions underpinning the organization and structure of education and pedagogy along with their resulting dualisms: human versus animal, culture *versus* nature and reason *versus* affectivity. We agree, for example, that what technology does is not to produce something unnatural (Ortega y Gasset 2007), but to probe the new possibilities that nature offers. Technology thus reveals that *relationships are constitutive*. This tendency is not typically human; it pertains to the phenomenology of life. Posthumanism, therefore, tends to reconnect what, for the sake of its anthropocentric program, humanism has separated.

The connective principle of posthumanism does not deny technoscience its importance. It actually frees technoscience from the anthropocentric shackles that, consistently with an a priori imperative, restrict its expression to a human celebration, or rather, the celebration of a particular idea of the human. I would like to underline the importance of this point, because anthropocentrism is ultimately a prison for the human, as suggested by the secondary role that the affective dimension plays in it. Posthumanist critique targets anthropocentric prejudices with repercussions on ethical and political levels (Braidotti 2017). Posthumanism is responsible for a revolution in bioethics based on four fundamental principles: (i) it is necessary to overcome traditional separations; (ii) it is impossible to base ethics exclusively on rationality; (iii) it is incorrect to think of the humanities in opposition to the natural sciences and to separate the descriptive from the prescriptive; and (iv) to confer moral ownership on humans alone has no basis, and is detrimental to humans themselves. Let us start from the first point, which, for example, assumes a separation between human and environmental ethics. We refer to the two worlds as if they were separate. In fact, today like never before, it is clear how tragically and mutually connected they are. Bioethics should have an ecological orientation and be based on the principles of relational systemics and complexity theory. With regard to the second point, we should avoid basing all our ethical arguments on rationality and deontological precepts (as Kant did). Ethics should be founded on human characteristics, including affectivity in its emotional and motivational components. Affectivity has an important influence on moral behavior (Spinoza 1951). Besides traditional prescriptive norms, therefore, we also need a *new sentimental education*. With regard to the third point, it is anachronistic and inconsistent to maintain the separation between the human and the natural sciences because it creates a parallelism between descriptive and prescriptive disciplines. An ethical system founded on the idea of creating a bridge between science and ethics (van Potter 1971) or based on operational responsibility (Jonas 1985) is no longer sufficient to build a constructive and consistent dialectic. We need a transdisciplinary and mutually problematic relationship, so that science may pose problems to ethics and vice versa. Finally, anthropocentrism remains the most serious impediment to the progression of a coherent ethical discourse, free from prejudicial circularities due to a priori prejudices. It is an error to consider the human the sole moral patient, but, most of all, the human cannot be viewed as separated from the world.

Posthumanism also takes account of the transformations in the ontopoietic interpretation of human predicates in ethical terms. The fact that our predicates are no longer viewed in essentialist terms means that we should be aware that our condition has always been *over the human* (Nietzsche 1917), that is, beyond its phylogenetic expression. The fact that human predicates do not derive from a process of either emanation or compensation has become all the more apparent with today's acceleration of infiltrative technologies. According to posthumanism, we need an ethical reflection upon the impact of technoscientific applications on life. The classic reflection on the relationship between what is technically feasible and what is allowed must be adjusted with: (i) a different view of the relationship between science and ethics: ethics must encourage rather than simply censor research; (ii) a metamorphic and hybrid interpretation of entities as relational and connected, rather than static and orbiting around an essence; and (iii) an anticipation of the possible changes that will be caused, firstly, by the ecological and epidemiological crises of our time, and, secondly, by the increasingly complex, systemic and immersive development of the technosphere. This dimension of the technosphere risks widening the gap between the human and the biosphere. It also risks enhancing our solipsistic and self-referential tendencies, to the detriment of our empathic capacities (Mortari 2015).

Through grafts of cybernetic interfaces and by virtue of traceability and biotechnological cosmetics, we are witnessing an increasingly frenetic evolution of technologies that massively affect the dynamics of the living. Thanks to the spread of the digitalization of media and machines over the last twenty years, we have also witnessed enormous transformations in our lifestyles. All of this raises the obvious question of whether bioethics, as it developed during the 1970s and 1980s, still has a role in our society (Prodomo 2020). While it is clear that, if anything, the role of bioethics has been emphasized by our current problems, it is also undeniable that today's picture is totally different. We cannot separate the different domains, as proven by our relationship with the biosphere. Indeed, the COVID-19 pandemic has exposed the fragility of a social and production system that does not take the ecological balance into account, and combines intensive farming with the destruction of forests to create space for monocultures (Adams 2015). Animals are abused in both livestock and marketing processes. Moreover, they are commodified in industrial animal husbandry. Human epidemiological exposure and animal abuse share the same preconditions and cannot be problematized in separate contexts.

The analytical approach that considers every single problem independently is not at all suited to understand the interactive dynamics that pervade the biosphere as a system. We cannot protect the human unless we take all the intervening factors of its life into account. The biosphere cannot be considered a home or a stage where the human can freely exercise its leadership role. In fact, the most suitable metaphor for the biosphere is that of an organism of which the human is an integral part. It is obviously impossible to protect an organ if the whole organism is degenerating. But it is also incorrect to consider the biosphere a stock of resources (Moore 2016), because it is actually a systemic entity rather than an aggregate of entities that can be quantified and used like a pile of coal. Any alteration produces chain reactions. This should not induce us to think that nature is intangible. Instead, it should urge us to adopt a systemic rather than an analytic perspective whenever we make a decision. The biosphere is an ecological entity, but this attribution has a semantic value that transcends the usual meaning of the term. Indeed, when I argue that, rather than an emanation, a predicate is the result of a relationship, I am applying ecological principles in philosophy. An ecological structure is not a habitat designed to favor the presence of certain species over others. At least, it is not only just this. "Ecological" means that the structure derives from the involvement of, and the mutual relationships between, different species. Hence, it is impossible to preserve that particular habitat without considering all the species that have co-constructed it.

The biosphere therefore has *different levels* to be taken into consideration, including the epidemiological one. Since every ecosystem has its own virosphere (Quammen 2012), it can become extremely dangerous whenever its homeodynamics are altered. Human bioethics must deal with ecology not only to preserve our living environment, but also to promote public health, which is also, always, *global health.* Similar arguments can also be raised with reference to the issues of environmental pollution and the destruction of the natural habitats of many territories. There is always an imbalance between the people who use the resources and those who are deprived of them (Gostin 2014). Ecological costs are never paid in equal measure by everybody and, in general, the poorest populations and the most disadvantaged social classes are those who suffer the most. We can ultimately conclude that the ecological question is of fundamental importance, and that a new form of exploitation based on these parameters now exists: making others pay for the costs of environmental destruction. Bioethics, therefore, cannot deal only with the problems of the beginning and end of life; it must also deal with all forms of exploitation of the human by the human that happen through ecological erosion.

Even the mistreatment of animals in intensive farms, which is linked to the consumption of foods of animal origin, is not merely a question of antispeciesism. This abuse is the main cause of deforestation, the production of greenhouse gases, the decrease in biodiversity, and the disarticulation of the most important biomes (Marchesini 1996). The consumption of foods of animal origin requires bioethics to address simultaneously issues related to public health and ecological dynamics in the biosphere. Posthumanist bioethics rejects separation and favors a systemic view of moral problems. Being systemic means not differentiating between a narrow morality based on abstention and a wide, proactive morality, based on ethics and virtue (Carr and Steutel 1999). The subject cannot hide behind disinterest and carelessness but must adopt an active role and become a guardian of moral issues. The subject must mature into someone working for the good as opposed to a mystic.

Similar arguments apply to the technosphere too. All technological innovations and inventions have repercussions for people that differ on the basis of parameters such as the wealth and psychological resilience of whomever is affected. We must bear in mind that technopoiesis is anything but egalitarian. Most of the time, it enhances differences by making the weak even weaker and the strong stronger. Each new technology accentuates the social divide (Norris 2001). As previously suggested, this is not only an economic divide. After having been introduced, a new technology usually requires time for people to learn how to use it. This period of time varies from subject to subject. At first, a new technology seduces the individual, thwarting his/her critical skills and making him/her easily manipulable. We could say, in short, that every new technology launch blows the alarm system of the self, leaving the inner safe open. This was true with the advent of the radio, television, the internet, and social media. The psychotropic effect of technomediation should not surprise us. The impact on people, however, differs according to many variables, including, for example, emotional fragility. This effect, moreover, tends to decrease over time. One of the duties of bioethics, therefore, will also be to understand what impact a particular technology may have on people.

## TWENTY-EIGHTH THESIS

Posthumanism is influenced by new artistic performances and proposes new aesthetic canons, seeing art as a form of experimentation with what is possible and as a form of knowledge that probes the meshes of reality.

We can argue that the crisis of the irenic vision of the human, who is standing at the center of the world as the measure of reality and represents the alpha and omega of all projections and projects, occurred throughout the twentieth century. This happened even before an approximate idea of the posthuman was drafted. I believe that philosophers such as Gilles Deleuze, Jacques Derrida, and Gilbert Simondon have to be credited with creating the theoretical preconditions that facilitated a connection between the critique of humanism conducted by Frederich Nietzsche and Martin Heidegger and posthumanist philosophy. The posthuman dawn was heralded by a change in the way of feeling, which reinstated the value of difference, of nature's vital force and of relational ontology. All these elements combined in the critique of anthropocentrism that was already being implied by Michel Foucault. The advent of digital technologies and biosynthesis would then clarify this metamorphosis in feeling and inspire artistic experimentations in the works of Matthew Barney, Orlan, Stelarc, Karin Andersen, and Daniel Lee (Marchesini 2010). Before emerging in philosophy, therefore, posthumanism was born in the workshops of artists, among new developing sensitivities. It freed the essential form, expressed itself as a reckless navigation across the seas of mutation and contamination, and experimented with various new existential dimensions created by transition, transience and hybridization. A new type of aesthetics came into being that was apparently provocative; but it would be more correct to relate it to a *new idea of the sublime*, based on the virtuality of *bios* and its form-generating power. Posthumanist aesthetics therefore emerged before posthumanist thought (Deitch 1992), even though much of the critical philosophy of the second half of the twentieth century had already created the conditions for this paradigmatic shift. The posthuman sublime cannot be related to traditional zoo/technomorphic aesthetics, even though it sometimes gets confused with it. It challenges static and predetermined forms and avoids anthropomorphic projections. The posthuman sublime can be described as a thrill and vertigo that does not originate from the overwhelming feeling of nature's original perfection, such as the one we can find in Romanticism, but from the poietic and Dionysian (unpredictable and connective) flow of *physis*.

As observed, a central figure in this change in sensitivity is the hybrid. It is an entity that challenges the attachment to a fixed form and encourages the transient and transitive (that is, transversal) character of the form itself (Dusi and Saba 2012). Posthuman art is a journey into knowledge. It cannot be considered a minor gnoseological experience (Baumgarten 1983). What emerges from it is a transreal view of the world. Posthuman art probes and even twists the real to allow new possibilities to emerge. In the last decades of the twentieth century, science fiction also contributed to drawing mutable scenarios related to possibilities of reshaping the body, altering memories, challenging identity stability, and introducing nonhuman perspectives and heterotopic and dystopic panoramas. End-of-the-millennium artists experienced new sensitivities as soon as they realized that the body has entered a stage of structural and functional fibrillation, expressing more and more visibly forthcoming metamorphoses. The new interest in the body differs significantly from humanistic panoramas, which display a *continuum* from Botticelli's hermeticism to *body art*. In the posthuman, the body is no longer an actor on the stage, it becomes the stage (Marchesini 2019). It is a hospitable body, wavering between the concept of ontological nomadism and that of multividuality. It is an organismic principle that continuously systematizes its relationships with the world and maintains a polyvocal character. This is noticeable in how the concept of alienation changes. This transformation is not a drift, and it does not express unease or ugliness. If, in the humanist tradition, alienation was a loss of authenticity due to external conditions, in posthumanism it means losing contact with the world, it means solipsism and the inability to join the *convivium*. We no longer speak of the inhuman, but of the overhuman. While this may suggest a kinship with Nietzsche's imagery, it differs from it because it is nonsequential and diachronically situated. In other words, for Nietzsche, the overhuman was a goal and an accomplishment. In posthumanism, it is a diachronic and permanently ongoing condition.

In posthuman art, metamorphosis expresses a shared feeling (we could call it "the unsaid"). Whatever its cause, this metamorphosis is not a descent into a degraded world, but it is not an ascension or an overcoming of limits, either. If anything, it is the exact opposite: conjugation into the everyday (there is nothing mythological about it), an adaptation to the underground and horizontality of contamination. We see the hybrid moving with extreme ease in anthropic spaces transformed into niches, not unlike a forest whose overlapping layers have unfolding and interweaving existential and living dimensions (as in the poetics of John Carpenter). Representations of the posthuman as cyborgs, theriomorphs, or other hybrid forms become extremely frequent in the subsequent decades and even lose singularity. It is the aesthetics of the ordinary mutant (Andersen 2006). The mutant exists in a daily panorama which, like in a memory store, mixes objects from different times: the antique stands next to the futuristic, as in the song *One More Kiss, Dear* by Vangelis in *Blade Runner*. Among the authors who portrayed this trans-chronic dimension of the everyday posthuman most explicitly, I would like to quote Karin Andersen and her *Insectoids* [*Insettoidi*]. The posthuman is neither shocking nor terrifying, and is not even considered an oddity anymore, so the crowd surrounding it is totally emotionless. The hybrid moves around by taking traditional means of transport, such as the subway, or it walks in the streets. But it can also take alternative routes, such as sewers. This all results in the copresence of a meta-anthropological biodiversity. Humans and hybrids live side by side, although not necessarily together. Regardless of whether they favor theriomorphism or cybermorphism, these artists' creativity transforms the body into a laboratory. The body becomes a canvas on which their imagination can run riot. However, it is no longer just the human who observes, as from a Benthamian panopticon. As happens with digital technologies, in posthuman art, the gaze is mutual.

The mutant's daily life brings us closer to a dimension of being that disrupts traditional canons. Change is traditionally perceived as the result of a before and after, a sequential process. Here, time becomes fluid: there are developmental steps but we stay inside a diachronic dimension. The metamorphic does not complete its transformation, its latent condition does not acquire a final shape or show a final result. The metamorphic cannot be compared to the transformations occurring in amphibians or insects, or to Kafkaesque creations or Apuleius' Golden Ass. The mutant character of the posthuman metamorphic is neither permanent nor accidental. It results from a contingency that is neither necessary nor suffered. Its traits do not say anything about monstrosity or duality (Fiedler 1993). They do not convey ideas such as deformity or deception and are not the consequences of traumatic events or *hybris*. They are the creative and—I would like to underline—unstable expression of ontopoiesis. Accustomed as we are to considering the human an essential entity who accomplishes ontogenesis by *ex-ducere*, the sudden appearance of an ontopoiesis upsets our interpretative points of reference. Ontopoiesis is not ontogenetic development, but ontological redefinition. Because canonical interpretations attempt to draw on the teratomorph, the *Doppelgänger*, the infected, the deformed, the brute, and the monstrous, they fail to grasp the truly innovative meaning of what stands in front of them.

The posthuman depicted by these artists simply reveals the mutant character of a human condition that is projected into the *pastfuture*. It does not express the canonical opposition between beauty and the ugly and repulsive, but it expresses the true aesthetic antinomy of the sublime. This interest in, and commitment to, the sublime shows that there is some proximity between posthumanism and transhumanism, so we can speak of a *contemporary sublime*, which is totally different from the Burkean sublime experienced by the Romantics (Burke 2020). However, differences are evident here too. Transhumanist imagery is dominated by cybermorphism and the postorganic body. Posthumanism instead is more directed toward nature's creative and relational potential, something that can build new worlds. I would like to stress the importance of this point. While drawing on the Heideggerian concept of creativity, posthumanism does not confine it to the human but, in line with the thinking of Nietzsche and Bergson, extends creativity to nature. Mutation is the main characteristic of an organic world (Braidotti 2002) that is continuously experimenting with new possibilities and actively operating in different living contexts by creating new existential forms. While ontocentrism is still present in transhumanism—we could actually call transhumanism an *anthropocentrism without the human*—what prevails in posthumanism is life in its plurality and mutagenic and relational commitment. The posthumanist sublime does not replace nature with *téchne*, but shows how *téchne* is an expression of nature (Arthur 2011). Moreover, the image we derive from transhumanist imagery is often transcendent or incorporeal, a state of near-fluctuation in a hyperuranion or on the Web: the backdrop is often empty and the dimension of conviviality absent. Posthumanism, on the contrary, underlines the context and reveals frequent infiltrations and the fact that it has a niche dimension. The technological apparatus cannot be thought of simply as a garment, because it changes the human somatic dimension. This reveals the dependences of the posthuman, its infections and adaptive correlations, its fragility and the extent to which it is exposed to its surroundings. Once again, we observe a profound difference between the transhumanist and posthumanist positions.

Posthumanism displays a powerful organic energy that emphasizes corporeity and the incarnated. The flesh, that is to say anatomy and physiology, is a workshop of experimentation. Posthumanism does not remove its attention from technological infiltration. However, it views the body as the real protagonist, even when it is subjected to the conjugations of technology. Flesh, with its sensitivity and creative capacity, expressed in morphopoiesis and physiopoiesis, ultimately dominates the scene. The posthuman body is never a postorganic body. It is, as it were, a hyperorganic body, whose augmented sensitivity derives from a celebration of the flesh (Deleuze and Guattari 1997), rather than an overcoming of it. In its representative operation, transhumanism is very simple: it replaces the Vitruvian Man with the robot, artificial intelligence, and the cyborg. No paradigmatic metamorphosis is needed, just a change in operators. Transhumanist imagery, therefore, communicates much more simply and easily than posthumanism, with its review program of ontocentrism. The posthuman of transhumanism is only projected into the future. Instead, posthumanism has a far-reaching historical amplitude. It is called upon to interpret the present and the past in the light of possible future scenarios. Admittedly, there are references to science fiction in both proposals, since we perceive we are in a period of strong transition. In our age, hopes and fears merge in a condition of permanent *thaumazein*, a state of wonder that nurtures our imagination, but it also makes us tremble because of the enormity of the events taking place. There is a huge gap between posthumanism and transhumanism in terms of both their cultural and philosophical backgrounds and their sentimental and artistic sensitivities.

Focusing on the human biological condition is typical of the posthumanist approach. As previously suggested, the strong influence of Darwinian thought on the posthuman cultural climate transcends the domain of evolutionary biology. Darwin started a laboratory of thought that continues to work intensely in order to undermine residual prejudices—or, rather, obstacles—linked to: essentialism, interpretative anthropomorphism, natural theology as well as top-down conceptions of nature that consider it a project or a backdrop and resource at the human's disposal (Mayr 1985). These are aspects of Darwinian thought that, in spite of their important phenomenological repercussions, are not always examined. After the Copernican Revolution the need to have a stable image of nature became very strong. Darwin made it impossible. I would like to mention just three of the fundamental changes brought about by Darwin's work: (i) nature is no longer a created but a creating entity; (ii) nature is a metastable entity that is continuously transforming, rather than a static entity keeping its original perfection; and (iii) mutation is the nominalist expression of a populational view of the entity rather than its estrangement from an essential form. Darwin's work causes a shift from a top-down view of nature as a divine project to a bottom-up view of forms as emergencies. Deprived of its center of gravity, the human has been drifting away, living in a permanent oneiric dimension, made up of dreams and nightmares. What we need to do instead is use this crisis as an opportunity to move forward.

Posthumanism is mainly the expression of a continental philosophy that had to come to terms with existentialism, the phenomenology of life, and poststructuralism. Its proposal revolves around: the critique of humanism, a systemic and holistic approach (e.g., in gestalt psychology), the search for a sense of existence and biopolitics. On the contrary, transhumanism can be considered a new form of Enlightenment with a strong analytical and empirical approach (Fukuyama 2002). The artistic difference between the two movements reflects different cultural climates, considering that posthumanism devotes almost an obsessive attention to the flesh and approaches technology with a critical focal lens, while transhumanism does the exact opposite. Posthumanism takes a deep interest in animality as a metapredicative condition, which can transfer from one species to another, and in ecology, as a close relationship between the subject and its niche. The results of transhumanism and posthumanism therefore cannot converge other than in a general projection into a changing future, where the human phenomenon is represented very differently from how it used to be. Both of them often focus excessively on shocking and uncanny effects, making every narration transreal and every representation cartoonlike. Yet the reasons for this could not be more different: while transhumanism indicates a process in which anthropocentrism is the absolute protagonist, posthumanism stresses that what the future has in store is an era in which anthropocentrism will be scaled down.

The change of anthropological perspective between posthumanism and transhumanism is made explicit, for example, by the theriomorphic artistic movement (Rignani 2014). Posthumanism does not deny human dignity but modifies it by relinquishing the humanist aspiration for emancipation and elevation. Posthumanism adopts a horizontal, copulative and participatory dimension, based on contamination and acceptance. It gets beyond both the narcissistic and hedonistic idea of the centrality of the individual, and the hyperhumanist idea of empowerment and salvation that we find in transhumanism. In the posthumanist vision, and, accordingly, in its artistic experimentations, the human is always located—never floating in space or uprooted on the Web. In other words, the posthuman becomes the image of a relational ontology of which the human eventually becomes conscious. Relationship is vital: it is what shapes the human and prevents alienation, namely loss of contact with the world and self-exclusion from conviviality. The resulting image does not convey the idea of a progressing drift that might end in a ruinous self-withdrawal. The resulting image discloses how rooted we are in nature. Of course, the posthuman is constantly moving and migrating, continuously mutating and redefining itself. But it is not detached from its surroundings and it does not abdicate the *convivium*. Rooting is not topological but relational.

These paradigmatic characteristics profoundly affect posthumanist sensitivity, which never abandons its eco-ontological foundations. Posthumanism questions the principle of individuation of the classical tradition by considering the subject a *holobiont* (Margulis 1991) rather than an individual. Therefore, dedication to the care and reception of otherness is considered to be the true purpose of humanity. When we examine how different the artistic production of posthumanism and transhumanism are, we cannot fail to notice that one exhibits an almost shamanic possession and projection into otherness, while the other, in its presumption of perfection, betrays distance and closure. Posthumanism underlines the importance of the concept of *mimesis* as an ontologically relevant moment in the human dimension. *Mimesis* suggests that, rather than on appropriation, the human dimension is founded on dialogue and hybridization as well as identification and acceptance. But we need to clarify the concept of identity to avoid the risk of considering posthumanist experimentation an ex nihilo construction that denies the existence of a human nature. To state that nature is creative does not mean that there is no legacy. Creativity is never a creation from scratch but always an innovative use of previous acquisitions (Piaget 1970). This shows again how difficult it is to understand the Darwinian principle of heredity with mutation: the two terms—inheritance and mutation—are not antithetical but mutually supportive. There could be no experimentation without a supporting history. Therefore, we speak of a plastic identity, which has a range of virtuality, and not of a liquid identity, characterized by totipotency. Recognizability, therefore, is not to be found in the anthropomorphism of the image—the robot in human form—but in the metacondition of animality with a precise phylogenetic conjugation.

## TWENTY-NINTH THESIS

One of the aspects that most differentiates posthumanism from transhumanism is its rejection of all forms of reductionism while embracing the theses of the positional logic of the entity and of the relational construction of qualities.

The twentieth century can be considered the age of the decline of humanism because of its ambivalent and ambiguous character that combined forms of anti-humanism with attempts to go beyond humanism by emphasizing anthropocentrism. In relation to the latter trend, the past century witnessed an increasing desire to control the phenomenology of life in both predictive and descriptive ways (Monod 1972), and in interpretative and prescriptive terms. Despite the development of complexity theory and the impact made by the gestaltian holistic view and chaos theory, the so-called short century arguably witnessed a prevalence of determinism. This was the consequence of a tendency toward reductionism. Determinism, reductionism, and control go together: they are all in line with anthropocentrism, whose long tradition precedes humanism but finds its full realization in it. The twentieth century, however, also marked the beginning of a transition that undermined the very presuppositions of this mathematization of reality (Stengers and Prigogine 1979) and, most of all, of the logic of identity and necessity. We have noticed, for example, that the advent of digitalization has not only paved the way for new technologies but has also profoundly changed our way of thinking, especially from an epistemological point of view. Let us consider three aspects of digital epistemology: (i) the idea that the form/manifestation of reality is not given or owned (authentic and self-referred), but develops through an *event of discretion*, the process that divides a *continuum* into discrete entities (Wheeler and Ford 1998); (ii) the idea that instead of an ontology of identity, we should speak of an *ontology of possibilities*, in which the entity is never given in a single form but is always intrinsically virtual (Badiou 2006); and (iii) the idea that an entity's predicates are not essentialist and autonomous expressions but follow a *logic of positionality* and they can be related to their relationships (Von Bayer 2004).

Digitalization is changing how we look at reality. This is happening not because our immersion in so-called virtual reality is pushing us toward unreality, but because it urges us to look at reality differently. Many of the assumptions of humanist philosophy are in crisis, such as a metaphysics based on individuation, necessity, the principle of noncontradiction, the universal principle, essentialism, the mathematization of the world and determinism. The case of new evolutionary biology is paradigmatic, because: (i) it views a species' niche as a construction (Lewontin 1983), namely a selection made by the species among a *mare magnum* of possibilities rather than the occupation of a predetermined space (the idea of an empty niche); (ii) it develops the epigenetic idea of the phenotype (Morgan et al. 1999) where the individual is the major operator of the forms that radiate onto the genotype and interrupt unidirectionality (from gene to protein), which was the dogma of neo-Darwinism (Wallace 1870); and (iii) the symbiontic view of the phenotype (West and Jane 2003) assumes that the organism develops a morpho-function on the basis of previously established relationships. This means that, albeit within a *range of conjugability*, the genome resembles a virtual entity, with more phenotypic possibilities that depend on external cofactors. It is the positional idea of the predicate that brings in the dialogical view of the entity, which does not come into being as such, but through relationships. This is what I have coined *heteronomy*: the fact that we assume the need for a relationship.

As we have repeatedly observed, this causes a profound shift in anthropology. In a relational view, the degree of virtuality, the range of possible conjugations, defines the positional plasticity of the entity and, accordingly, its phenomenological productivity. The range of possible conjugations is determined by the complexity of the system, in other words by its redundancy and not by its deficiencies (Johnson 2002). Let us take an example: it is by virtue of our excess of neurons and synapses and thanks to the complexity of our neuromodulation pathways that we can experience multiple cognitive states and biographical configurations (Edelman 1987). A redundant system amplifies positionality because it creates a very large space for conjugations—that is, a variety of possible positional states. This makes it particularly susceptible to the relationships it establishes. This is not due to a deficit of content, but to content redundance. Moreover, the entity is no longer a collection of predicates, but *predicative openness*. In other words, through relationships, the entity grants itself the possibility of letting new predicates emerge. This is what we mean when we speak of a philosophy of gift: it is a moral prescription or a prescription of values as well as the first principle of ontopoiesis (Marion 2018). I have repeated several times that posthumanism is a philosophy of *constitutive relationships* rather than of interaction or dialectics. Even considering culture the fruit of the human's relationship with the world is a step toward a predicative reinterpretation of the entity. The concept of epiphany should not be trivialized and taken for the mere imitation of animal and vegetable performances. It points at the positionality that the human adopts toward otherness—for example, feeling like a bird. Projecting oneself into otherness means imagining their condition, relating and, consequently, hybridizing with them, hence experiencing—in part at least—how it feels to be a bird (Marchesini 2017).

Reality is a *continuum*. However, if I want to observe it, I have to break this continuity and define discrete entities. If we consider the entity an expression of discretion—accidental or probabilistic, perspective or experimental—we do not deny its *continuum*, but we understand the logic of its manifestation, which must contemplate some breaking points or points of salience. In quantum mechanics, this is the energy package, in genomics the gene, in the river its bed, and in the individual the continuity of his/her history. The bit can express itself in different ways, for example in a *threshold system* that operates with an fb– below a certain gradient (damping perturbations and producing homeostasis) and an fb+ above it (amplifying perturbations and producing autocatalysis). Many hyperobjects express their identity by adopting a threshold configuration. Their identity manifests itself in a state of dynamic equilibrium but also in the possibility of an evolution—as with the carbon cycle. Examples of threshold systems include the positioning of an electron on a certain orbital, catastrophes, weather trends and pathologies. A threshold system is, by definition, relational; not because it lacks an inherence, but because it is predicatively open, it does not emanate predefined predicates or is self-referred. Therefore, the philosophy of the digital epoch (Waldrop 1992) can be considered the epistemological background that somehow unifies—or, at least, helps us understand—theories that developed throughout the twentieth century, such as complexity theory. These theories share a common ground in the principle of potentiality and positionality. This principle challenged the interpretative paradigm that used to draw on the concept of extension to determine the entity by mathematical reductionism (Longo 2021). The digital dimension shows the force of discretion and induces us to consider the phenomenological singularity of positionality. But, because of this, we need to go beyond the discrete manifestation of the phenomenon to understand the principle of possibility in the real.

The contradictions of our time are all in here: on the one hand, we have adopted digital technologies as tools that enhance our performance possibilities and provide us with particular services; on the other, modernity's philosophical criteria and points of reference are still being eroded—much more radically than we think. We are prey to the hyper-humanistic temptation of considering our age an acceleration of the need to celebrate the human and create a world on a human scale. Hence, we either invoke a kind of New Renaissance (Borgato et al. 2010) or simply gather conceptual resources to preserve an anthropocentric vision. Human supremacy finds in the enormous possibilities offered by digital technologies new arguments to support human superiority over nature. We return to Descartes via the model of neurosimulation and annihilate nature through immersion in virtual reality made possible by digital technologies. In this sense, we can say that, because digitalization facilitates our progressive divorce from the biosphere through self-referential withdrawal, humanism is reinstated and accelerated. Most of the time, hyperhumanism turns out to be mere conservatism, avoiding coming to terms with the theoretical consequences of digital philosophy (Longo 2013). We have observed that many thinkers are not content with inaugurating a new Renaissance era. By drawing on the emancipation principles of the Enlightenment, they further dream of a new ontological dimension that can no longer be considered properly human (Savulescu and Bostrom 2009). We find ourselves within the vast transhumanist panorama, which considers technology a means of accessing new existential dimensions and getting rid of the fragility of the organic. Posthumanism—the philosophy of *bios*'s creative vitality—adopts the opposite approach: it takes digital epistemology to its extreme consequences in order to reveal the virtual condition of the real and its positional phenomenology.

We find ourselves within this contradiction, but there is no need to be afraid, provided we can appreciate the relational productivity of points of view. Let us take an example. When we state that viewing the human as deficient is pointless if we have to understand technopoiesis, we do not deny the principle of dependence in absolute terms. In fact, the principle of dependence is already implied in heteronomy and we simply reposition deficiency from a conceptual viewpoint. Living inside contradiction means that we accept a plurality of points of view, including our own. Each statement can, in fact, be framed differently depending on its positioning. This does not mean that we annihilate it or deprive it from meaning but that we can understand it better because we know where it is positioned (Putnam 1999). Posthumanism does not deny the overhuman but simply places it in a logic of *continuous multidirectional transformations* that does not imply going beyond a particular condition but opening up toward new conditions. According to posthumanism, therefore, the idea of nature as extension derives from having favored a particular epistemological discretion: it is not wrong, it is just limited and insufficient, namely singular and related to a particular perspective event, which can also (though not only) be referred to the biography of that particular philosopher (Simmel 1977). The error, therefore, lies in wanting to prioritize a particular point of view by removing it from its perspective framework. But beware: this is not an absolute relativism. Every perspective frames only one possibility of the real, one of the conjugative spaces implied by inherence (Ferraris 2012), so every single perspective only brings out the possible profiles. Certain aspects of the real can only be understood by refutation and when the forms we would like to confer to the real find no room for possibility. The critique against humanism addresses its absolute view of the human, the purpose of which is to portray the human as a privileged and autonomous entity. This critique does not deny human peculiarities, that is, it does not assimilate the human with the other species. It reveals that human specificity consists of a particular conjugation of animality; it is placed within animality, rather than in opposition to it.

Considering the digital metamorphosis merely an acquisition of tools needed to achieve particular performances or obtain particular services does not help us take the paradigmatic leap encouraged by posthumanist philosophy. The latter's purpose is to relativize humanist thought, which is a perspective that transforms anthropomorphism into anthropocentrism. The inability to fully accept the philosophical consequences of digitalization—that is, the positional logic of predication—means being unable to interpret the vast and complex challenges of our time. Posthumanism seeks to raise our awareness of the narrowness of the humanistic paradigm and the risk underlying classic and transhumanist hyperhumanism, being their inability to reveal what panorama new technologies are putting in front of us. The nuanced reference to Cartesian thought that permeates hyperhumanism, with its tendency to view life merely as extension not only reduces our ability to understand ourselves, since we are an expression of life too, but also thwarts our imagery. Questioning the consistency of the real—the idea of the world as imagination—is a *leitmotif* that deeply influences science fiction: let us just consider the narrative *topos* of neurosimulation (Irwin 2002). However, annihilating the real is only just one of the consequences of humanism. Neurosimulation, therefore, is nothing but the use of digital technologies to enact Descartes's ramblings in a simulated world and it does not help reveal the epistemological consequences of the uses of digital technologies. Putting on hold the real is a humanist ploy to enhance our autonomy, exactly the opposite of the entity's positional logic (Nancy 2000).

When I argue that, in spite of its sci-fi ramblings, transhumanism is a retro thought, I am referring to its tendency to use technology not as an epistemological lockpick to enlarge our perspective of the real, but as a tool to realize old humanist aspirations—or rather, old humanist dreams. Scenarios displaying a human who is increasingly withdrawn into itself and surrounded by nothing are the projects and perspectives of transhumanism, for which liberation means the acquisition of self-determination (Kurzweil 2005). Even though they are scientifically untenable, we need to examine the different expressions of twentieth-century reductionism and understand how they influence the transhumanist imagination. This is important because, while hyperhumanism fantasizes by drawing inspiration from a technology of which it has no knowledge and is still considered in terms of a "tool-useful-for" something, the consequences of the technological revolution transcend our ability to interpret them due to the very functional approach we are adopting. As the gap between science fiction and technoscience shows, technological imagery is lagging behind the development of technology. In just a few decades, the two swapped roles: up to about eighty years ago, science fiction drew inspiration from technoscience; since the middle of the last century, the opposite has happened. This means that, in the past, science fiction was the vanguard of *téchne*; today, it is actually its rearguard. This is the reason why most of the transhumanists' fantasies are resoundingly disproved by scientists. Their appeal prevails in cultural areas that, regardless of the solid funding they receive, show scarce scientific knowledge. The scientific illiteracy that is widespread among politicians as well as across philosophical research areas with a predominantly humanistic approach is what helps to focus on transhumanism. Instead, a systemic approach should be adopted.

My criticism of transhumanism also addresses its remoteness from the phenomenology of life as well as the scarce attention it devotes to, and I daresay, its scarce knowledge of, the biological phenomenon as a whole. Posthumanism, on the contrary, fosters the body, corporeity and the living, and considers *téchne*a biological expression rather than a separate or even oppositional dimension (Popitz 2017). All expressions of the real are positional, although the phenomenon of life is undoubtedly the most relational we can imagine—or the one that is dependent on relationships the most. At a biochemical level only, the living builds discretions based on iterations and redundancies that may develop enormous possibilities of conjugation. For instance, the fact of having billions of neurons is a redundance that enables us to acquire many different individualities. Not to mention that, from a thermodynamic and evolutionary viewpoint, living systems can never be said to be autonomous. They depend on external contributions: synchronic ones, such as the sunlight for a plant, and diachronic ones, such as an inherited phylogenetic legacy. One of the major errors of humanism is its reductionist interpretation of the living (Polkinghorne 1987), viewed merely as a repetitive and adaptive mechanism. The living is actually a *creator of worlds* that, by positioning itself relationally, is continuously building its existential singularity. By doing so, the living enacts one of its possibilities and leaves a trace of it. All this is as much in line with digital epistemology as can be. It is in analyzing the phenomenology of life that the gap between the post-humanist and hyper-humanist approaches becomes evident. The latter seeks to promote a transcendent human by belittling the contents of the rest of the living. Reductionism is, therefore, one of the characteristics of the humanist presumption to standardize bios as extension by extracting the human out of the flux of continuity. This is because the human is conceived as *res cogitans* contained inside a body. In many respects, the twentieth century was the ultimate attempt to create a bulwark that would be able to defend the integrity of a humanism in deep crisis. With *The Origin of Species*, in fact, anthropocentrism had suffered a hard blow that would influence the thought of the greatest philosophers of the late nineteenth century. In order to reduce the phenomenon of life to sheer extension and enact something like a counter-reform to Darwinian continuism, using Occam's razor as an axe turned out to be an immediate and urgent necessity.

Neo-Darwinism was one of the first forms of reductionism. It drastically reduced the complexity of Charles Darwin's reasoning about the engine of evolution (Pigliucci and Müller 2010) and embraced the vision of Alfred Russel Wallace, which similarly reduced the complexity of the transformation mechanisms in the species. Natural selection became the sole chiseler of a species' traits. Moreover, a kind of tautological personification was attributed to it (adaptation as the survival of the fittest). The emergence of forms, on the other hand, was considered the result of pure chance. This interpretation thrived from neo-Darwinism to sociobiology, and the living was denied an active role in adaptive processes (Wilson 1975). Species became passive entities, subject to the continuous and gradual work of natural selection. Each character had to be explained and mathematically accounted for on the basis of the utilitarian principle. There is a profound link between neo-Darwinism and genocentrism, namely the idea that the gene is responsible for the traits as well as all the characteristics of the phenotype. Significantly, the sociobiological view brought together both assumptions and formulated bizarre hypotheses about altruism in nature and social collaboration, such as *kin selection* (Hamilton 1964). Genocentric reductionism was also welcome in peer-reviewed publications because of a misguided observance of Occam's razor. Yet the deterministic view turned out to be completely incorrect: the gene certainly stores the information that enables protein synthesis, but the translation of the genome into the phenotype is ruled by nongenetic markers and the presence of cytoplasmic and environmental activators. To persevere in a view of genetic determinism, which presumes to identify the gene for each single orientation of the living, means sticking to an already refuted hypothesis. After all, the twentieth century experienced all the contradictions of a twilight period: starting from the 1970s, evolutionism would be revitalized by authors such as Richard Lewontin and Stephen Jay Gould. They would focus on the historicity of the species' evolution, the complexity of phenomena, the alternation of long periods of stasis to sudden changes, and the groundlessness of the Panglossian justification of characters. But the final blow to neo-Darwinian reductionism was delivered by epigenetics, the *evo-devo* proposal, and the theory of niche construction. These three interpretative points of reference—integrated today in the so-called "extended theory"—have undermined for good both reductionism and a passive conception of the living. The algorithmic idea of evolution (Dennett 1995) has thus been transcended once and for all.

Reductionism also played a part in our view of animality and became a kind of counterreform to the continuism of Darwin's epigones (Romanes 1892). Lloyd Morgan's canon of parsimony was adopted to deny a mental condition to other species and transform them into entities driven by automatisms—in spite of Morgan himself. There is obviously no difference between saying that animals are nothing but machines, as René Descartes did, and that they are driven by automatisms, as in the tropistic, pulsional, or associative theories of the first decades of the last century. In this case, reductionism leads to expressive determinism: the two schools that, in the first decades of the twentieth century, contended with each other to gain primacy in the explanation of animal behavior were North American behaviorism and Central European psychoenergetics. For both of them the mentalist hypothesis was a sort of taboo, and the expression of behavior was a form of mechanism (Griffin 2001). Behaviorism refers to a form of environmental determinism: external stimuli cause reactions that produce a contingent response and build conditioning via reinforcement. On the other hand, psychoenergetics refers to the idea of an intrinsic drive, namely to an interpretation based on hereditary determinism, which in short becomes genetic determinism, and assumes that animals are driven by instincts. Reductionism apart, we notice how these explanations hypothesize a repetitive entity without creativity, completely confined—Heideggerianly stunned—within its own behavior when interacting with the world (Heidegger 1992). The mechanistic view does not contemplate a presence or, therefore, a positioning of the animal in a here and now. This is because its behavioral expression is always the echo of something that happened in the past, an instinctual predisposition or the result of conditioning. In addition, the animal is not considered a subject because it is driven by automatisms and has no self-ownership. Viewing behavior as the expression of disjointed and self-referential automatisms rather than the fruit of a systemic is as far away from what we have defined as a "digital epistemology" as can be. The analytic view is based on the summation of parts, whereas the systemic view is based on the interaction among parts. In the former, we always obtain the same result; in the latter the results are different, because the outcome depends on how the parts are organized within the system and, consequently, on how they interact. The analytic view assumes the possibility of disassembling and isolating the system's parts. This strips the system of virtuality and positionality—both of which ensure the possibility of multiple results, as is observable in animal behavior.

Under the guise of systemic considerations, cognitivism is a more insidious form of reductionism. Cognitivism is the idea that the mind is nothing but a packet of information regulated by elaborative mechanisms (Johnson-Laird 1988). Cognitivism is the extreme expression of the theory of the inner processor. As an approximate metaphor of how information is managed, it is certainly a step forward in comparison to response theory, based on the stimulus-response mechanism. Problems start when we switch from the metaphor, or explanatory heuristics, to analogy and make an overlap between the brain and the computer. This leads us to believe that our neurobiological apparatus is a kind of hardware, running a packet of information that defines our mental experiences (Lindsay 2013): desires, emotions, knowledge, memories, and decisions. Such an interpretation has been widely challenged by the most important neurobiologists. Yet it has nurtured science fiction, especially that inspired by Philip K. Dick and cyberpunk that fueled the transhumanist imagination. Among the ramblings of mind-downloading, the implantation of memories or knowledge, backup copies of memory, transcendence in the network, and the translation of identity into pure information, we have created a post-organic imagery from which the transhumanist proposal can draw inspiration. We have depicted scenarios of existential heterotopias that have one big problem: they are based on an incorrect assumption. Our brain is not a computer, nor is our mind a packet of information.

The hypothesis of the brain in a tank that transforms the body into a set of stimulative and operative devices that are replaceable to become more functional, resistant, and interchangeable is yet another distortion similar to reductionism: it is the idea of the modular body. What we are facing here is the most obvious and, in my opinion, disarming aberration, because it is so banal. The mind is the expression of the body, of which the synaptic connectome is part. But it is not the sole element responsible for our psychological state (Damasio 2003). Moreover, the functional complexity of the organismic systemic structure, with its possibilities of synergy, plasticity and replacement, is the fruit of millions of years of biological structuring, both in terms of setup and performative endurance. Should we replace such an extraordinary structural and organizational work with modular equipment? It would be ridiculous if it were not so tragic. Reductionism reminds us of the Cartesian appeal to dualism. The operation is very simple: transforming transcendental *res cogitans* into *res informatica*, a process that reveals the intent and goal of transhumanism: to convert Pico's transcendence into immanence. The objections we raised against reductionism and its advocates all help us understand the posthumanist attitude, which considers the phenomenology of life in its full complexity (Merleau-Ponty 2003). Posthumanism knows how to accurately differentiate between sci-fi imagination and technoscience, and rejects the idea of an overlap between science and transhumanism, because there is none. Complexity theory adheres to biological phenomena and tries to interpret and, if anything, direct them, though not in such a way as to have absolute control over them. Both determinism and reductionism are rejected because there are layers of reality. We posit, for example, that the rules of physics and chemistry are the foundations of existence, yet that species have their own emerging characteristics, so that their phenomenology cannot be reduced to physics and chemistry (Lorenz 1973).

## THIRTIETH THESIS

Posthumanism questions the substantive concept of reality, as defined by a specific form, considering it rather as a set of layers, each endowed with a range of virtuality from which to extract possible forms.

While humanism has led to forms of annihilation of all the expressions of the real, and has attempted to absorb them into the human dimension (Žižek 2002), with posthumanist philosophy, we can exclaim: "Welcome back reality!" We have underlined how digital epistemology shows that predicative emergence results from events of discretion, events that interrupt the *continuum* and let discrete forms emerge. Reality is, therefore, a virtual structure whose possible configurations are defined by the individual perspectives that dialogue creates (Austin 1962). The form acquired by the real depends on the type of questions asked. Let us take the example of a human and a dog. Their dimensions of reality are very similar because both are part of the biosphere and share the same layer of reality, and their temporal and spatial dimensions are not those of electrons or galaxies. Nonetheless, the configurations that emerge from them are different, because their perspective structures—their sensory and cognitive structures—are different. They do not invent the panorama that emerges through them but manifest it; this panorama is one of the expressions the real may acquire. These expressions reveal something that *can be* in the real. There is a coherence between their perceptive structures and the real for two reasons: (i) ontologically, because it is the manifestation of the real that has shaped their perceptive organs; and (ii) epistemologically, because the emerging configuration is one of the possible conjugations in the field of possibilities of the real. Let us not confuse digital epistemology, which is based on potentiality and positionality, with absolute relativism or the negation of reality. In fact, we need to dig even deeper into the abyss of reality in order to allow the emergence of not only what is *already-given* but of possibilities too (Waldrop 1992). It is worth repeating that no predicative statement is the emanative expression of an essence. Reality is the set of the possible relationships that define the horizon of a subject's nomadic ontology. At the same time, however, it is also the matrix that brings to the fore the ontological importance of relationships. This points to a direction that I consider high priority, especially in the current ecological crisis: we *are* because we are *in a relationship* with the world. Destruction of the world is inevitably self-destruction.

There is a common thread between ontology, epistemology, and ethics. Therefore, the change brought about by posthumanist philosophy is prescriptive in being descriptive and the other way around. Rediscovering our connection with the world, that *convivium* with the biosphere that means participation and giving, is the way to go if we want to face the great challenges of our time. The posthuman indicates a transitional period in which existential experimentation will increase along with a feeling of precariousness and a need for certainty. The misleading idea that we need to go back to traditions in order to save ourselves from catastrophe will take hold of us and we will forget that the problems of today are the fruit of the mistakes of yesterday (Lanier 2013). Our horizon will very likely become fragmented and the human ontological dimension will begin to melt. We will certainly need to address challenges such as: (i) an unprecedented ecological crisis that will force us to look for solutions, also in terms of lifestyles, especially energy and food; (ii) a social crisis, especially in terms of integration and coexistence, due to massive migrations of people, but above all, the increasingly significant and unpredictable epidemiological events and the erosion of the planet's inhabitable areas; and (iii) a radical anthropological crisis produced by the advent of new technologies. With regard to this point, we need to observe that: (a) the rate at which technological development will accelerate cannot be predicted today; (b) new technologies have a spillover effect on the body and cause decentralization; and (c) the implementation of new technologies creates functions that, as we have witnessed in the case of the computer, only become visible in retrospect. The complexity of the factors at stake should make us conscious that transitions can always only be partially controlled, that the future is open and full of uncertainty. This is not something that should block us but rather prompt us to be cautious and adopt a broad perspective.

We need more science and scientific rigor, hence, we should abstain from sci-fi ramblings. A scientific approach is problematic and tends toward skepticism; it needs evidence: scientists rarely get carried away. All transhumanist projections into a postorganic dimension—uploading the mind onto silicon, the modular body, etc.—are extremely risky, and even the concept of artificial intelligence is far-fetched. How can we recognize intelligence? What is intelligence? And then, how can we find a definition of intelligence that is unique, univocal and non-anthropomorphic? We know that our empathic propensities, useful in many social and parental activities, can easily deceive us (e.g., we might anthropomorphize a robot and consider it endowed with subjectivity) and in this sense, the Turing test does not stand up to the facts. Ethology has disclosed the cognitive plurality of the different species and revealed that intelligence is a function, hence a specialization (Pinker 2014). The principle of individuation is invalidated by the impossibility of defining intelligence. Intelligence is indeed a blurry concept, extremely difficult to pin down. It can be collective, as in insects; relational, as in the relationship between mother and offspring in mammals, and it can be what holds together the symbionts that make up a body. The concept of intelligence is so hazy that the principle of individuation cannot encompass it. It is undoubtedly more correct to bring the meaning of intelligence back to its semantics: *intus legere* means to go deeper, namely to question and look for opportunities that go beyond those already experienced. In order to do this, however, you need to have inherent interests, an affective structure, which a machine does not yet have.

There is a striking difference between the posthuman attitude and the triumphal tone of transhumanism, for which technological enhancement is a metamorphosis from the human to the divine. Although transhumanism questions the principle of human perfection of the early humanists, it maintains their anthropocentric arrogance and aspiration to verticalization. Placing the human as a transitional entity between brutes and angels is where much of Western philosophy stumbles. We do not question the fact that technopoiesis modifies the space of human conjugation; we dispute that it provides mere enhancement. This is an irenic view because it neither upsets nor problematizes the human, and only offers the additional operational means to realize human dreams inside a fixed and stable view of human ontology. In the field of ontopoiesis, the posthumanist proposal is more radical and fluctuant: if the human changes, its dreams will also change. We can observe this also in the course of our lives: our dreams and aspirations change from childhood to adulthood. In other words, a shift in our existential dimension affects and modifies our goals. Changing the space of our possibilities means changing our entire ontological orbit, losing our barycenter and adopting a new one: it is not just dreams and goals that change, but the very meaning of *Being-there* (Heidegger 1927). Changing our space of conjugation does not mean acquiring more tools to achieve what today's human longs for, but gliding into new existential dimensions where new possibilities, unprecedented problems and different goals stand side by side.

Conceiving of technology as a tool that will liberate us means wearing rose-tinted spectacles: contemporaneity itself shows that every invention and innovation brings with it advantages and constraints. Besides, because every invention and innovation changes the order of factors—not only performative factors, as we have noticed—it allows for no going back. I am flabbergasted by how transhumanism conceives of the human's possibility of reaching the state of a god. What surprises me is not the arrogance but the stupidity of the idea. To analyze the changes in the conjugation field means to maintain a condition of *epoché* (Husserl 2012) and try to depict possible scenarios, preferably more than one, on the basis of what history has showed us. History—our first teacher in this question—has taught us that every single innovation has produced a change in human life. Every single innovation has had an effect on, or an interaction with, other technologies; it has enhanced certain techniques and completely changed human life, with repercussions on the human's critical abilities and emancipation. It is necessary to maintain a suspension of judgment, precisely because a technical innovation is never simply an enhancement but a change of all the subject's ontological points of reference and convivial dynamics (Nancy 2000).

What we are from an ontological viewpoint derives from a force field resulting from the set of predicative relationships we have established and the internal order we have reached, which defines a recognizable center of gravity we call human. It is important to understand the concept of *field of conjugation possibilities*, for which we need an adequate epistemological focal lens that can go beyond reductionism and determinism. Posthumanism adopts some of the assumptions of complexity theory (Kauffman 2019), such as: (i) the interactive effect that the predicates manifest in a complex system and that expresses itself in terms of synergies, antagonisms, and emergencies between its constituting factors; (ii) the importance of the information introjected by the system; and (iii) the quantum or threshold effect that creates layers of reality, for example in the transition from a chemical condition to a biological one. This means that, as with niche construction in evolutionary processes, the technopoietic insertion produces an alteration in the internal organization of the system and in its predicative relationships with the outside. As previously observed, this process is not directed and controlled by the human but it implies a change of the factors at stake. The process marks a transition whose lines are blurred and, above all, it defines an itinerary over which we have no control.

Apart from its philosophical limitations by comparison to the continental tradition, transhumanism betrays scant knowledge of Darwinism, physiology, ethology, or complexity theory. In particular, it seems to ignore the existence of organic layers (Plessner 2019). Posthumanism, on the other hand, takes a very different epistemological perspective and produces a paradigmatic change that is neither intuitive nor easy to understand. To move from a view of the real based on essentialism and the idea of the emanation of predicates to a relational or field-function view means hypothesizing different modes of predicative phenomenology, as many as there are possible fields, namely fields made possible by the network of interactions. This implies a phenomenology of the real that has no form in itself, but whose form depends on the particular organization defined by the relationship. Therefore, any change in the field is also a morphopoiesis (Rullani and De Toni 2019). Also from an epistemological viewpoint, a piece of technology is never simply the enhancement of a previous observational interface, but it causes an epistemic change, bringing out new predicates: the telescope and binoculars are not only enhancements of the eye, they also modify what we can see; they change the perspective and panorama we observe. The profound paradigmatic shift brought about by posthumanism becomes evident here: it rejects both an alleged objectification of reality and an annihilation of reality in the name of an absolute field of observation.

When we affirm the ontopoietic transition, we are within a logic of constraints and possibilities that somehow recalls the concept of evolvability of a biological structure (Kirschner and Gerhart 1998). In the end, this is what it is, although the niche is so emphasized here that it produces something qualitatively new. Posthumanism is not an exclusively anthropological discourse: it is about our relationship with the biosphere, about a comprehensive view of animality, about bioethics as a unified perspective and aesthetics as experimentation and research. One of the aspects that mostly characterizes posthumanism is the epistemological question about the definition of the real. As predicates are considered the expression of relationships, posthumanism rejects the possibility of reaching an ultimate view of reality and looking for a primordial substance of the real. Since any aspect of the real is the expression of an accomplished relational configuration, posthumanism considers any discovery about the real a morphopoiesis possible within a particular relational framework. This does not mean resetting reality but considering it a virtual entity that may take different forms and reach different layers of complexity, within a polyvocal conception of being.

Let us clarify this concept, which is crucial to understanding the transition. The problem of the reductionist vision is that it locates all expressions of the real in the same predicative layer so that the real appears to be endowed with an ultimate or univocal form whose quality is determined by the assemblage. The second option within the humanist paradigm represents, as it were, the other side of the coin: it assumes that reality is nothing but an impression and that it can take any possible form, depending on the observer or the observation method (Rorty 2017). Both views express a dualistic attitude that could be termed "disjunctive epistemology," and that is only partially corrected by the uncertainty principle and the acknowledgment that the observer modifies the observed. On closer inspection, however, we realize that they both are, more or less, concealed expressions of determinism. If we do not think of the predicate simply as substance, but as the expression of the relational moment, we realize that the real is a virtual condition, with both internal constraints and fields of possibilities. Accordingly, the epistemic process becomes a dialogue, that is, a predicative construction built on relationships, which has an irreducible asymmetry (Marchesini 2014): we cannot say what is true, but we can say what is false. I can only deduce what the constraints are, namely where the real does not permit relational predication, because all that is present and manifest is not the ultimate form of the real—just one of the many relational possibilities that the real makes possible through the expression of predicates.

From the latter point, the idea of dialogical epistemology arises. Every form derives from the type of dialogue established, and the real turns out to be univocal not in showing what is possible but in showing what is not (Godfrey-Smith 2021), namely what is not predicative in a relationship, because it is not envisaged by the virtuality range of the real. Hence, there is an asymmetry in knowledge. This asymmetry, however, is not due to the inadequacy of our instruments of knowledge, but to the fact that the real takes no ultimate form. The real is a condition of virtuality framed within certain constraints, which are the only thing I can know. This means we need to speak of layers of reality that are defined by the level of relationship we are considering because each layer is related to different predicative emergencies (e.g., the layer of atoms has a predicative structure that is different from objects, which have a different one compared to galaxies). Assuming the existence of a single form of the real means favoring one layer over others: we do not get incorrect feedback, but it is inevitably incomplete. This incompleteness is not due to the inadequacy of the investigating organ but to the impossibility of comprehending all the phenomenological layers at a single glance (Merleau-Ponty 2013). It is a mistake to think of reality as a measurable and stable form. Our presumption of being able to grasp the ultimate form of reality reveals the projection—arguably the anthropomorphic projection—that informs our processes of knowledge. Just as we cannot observe all the sides of a solid object because of its multiple layers, the structure of reality cannot be grasped at a glance: each layer unravels a particular dialogical relationship with its observer.

The dualism of counterparts is an optical illusion because observing means giving shape, namely, choosing one among many possible forms. This is a fundamental point to bear in mind if we want to avoid slipping into a form of neo-idealism. But since the observer can never accomplish more than one dialogical level at a time, the real will appear to him/her to be a stratigraphy of meanings: it would be arbitrary to subsume them in one level. The idea of the *layers of reality* is a very ancient one in philosophy (Berti 2010). Yet we had to wait for the last century's theorists of complexity to have emergentism and organizational information defined, which refer to how every single phenomenon is affected by the relationships among its constituting entities. Once more we observe that the real can express different qualities, namely qualities that are not substantialistic but dependent on its internal relationships and on the interactive level established with the observer. The higher the scale of complexity, the wider the range of emergence qualities along with the content and organization of internal information. Yet this system also depletes part of its virtuality with a subsequent increase in the constraints the observer experiences in the process of dialogical morphogenesis. In other words, the more complex the system, the better the observer can define what the system cannot be. At the same time, however, the more complex the system, the less it can be controlled.

Simplification is, therefore, the easiest way to control a phenomenon. By either transforming into noise or selecting a biased reference scale, we get the false impression that we enjoy full control over phenomena. The problem of epistemological anthropocentrism (Bachelard 2002) is not simply related to the obstacles of magic-projective thought or the so-called naive physics. Because of certain human tendencies and the cultural framework in which we are immersed, epistemological anthropocentrism is actually gaining ground and developing like a metastasis. These conditions risk convincing us that we are building new perspectives for the future while, in fact, we are just playing a transreality game. In other words, we want to understand the present—its ideas, difficulties, and aspirations—by changing the surrounding conditions. One of the greatest problems of anthropocentric epistemics is its tendency toward dualism and projection: by putting the subject center stage and emphasizing the discontinuity with otherness, anthropocentric epistemics cannot understand the system's evolution that comprehends the human. A penchant for disjunction and discontinuity creates gestaltic oscillations that resemble a process whereby an icon becomes visible by emerging from its background. In spite of what the term might suggest, transhumanism is not about transition but narcissistic fulfillment.

Posthumanism, on the contrary, denies both the idea of humanist perfection and of transhumanist fulfillment. Posthumanism is a philosophy of transition, viewed as a change in the predicative field of the human. It is the fruit of the joint collaboration of ecological and technological factors that are only partially controlled by the human through an intervention that is continuously in the making, changing all along. The human is an open construction site with no clear-cut project. It is certainly driven by our hopes, fears, and phylogenetic legacy, but it is also attracted and affected by the possibilities that surface along the way. The human is a journey made up of moments rather than destinations, where the memento of its sole final destination—one's own death—has no other real meaning than reminding us that life is a gift, a dissemination of works and days that make sense only when devoted to others. Posthumanism, then, is a search for possibilities, the acceptance of others as codesigners of our existence. Because of its stated refusal to define a center, it is a continuous shift in our center of gravity, the rejection of an essential form, of an ideological orthodoxy, and of a strong and stable identity. It is an appeal to pursue all initiatives in research and hybridization.

# Bibliography


Accarino, Bruno. *Antropocentrismo e post-umano*. Sesto San Giovanni: Mimesis, 2015.

Adams, Carol J. *The Sexual Politics of Meat*. New York: Bloomsbury Academic, 2015.

Afer, Publius Terentius. *The self-tormentor (Heautontimorumenos)*. Indianapolis: Bobbs-Merrill, 1963.

Agamben, Giorgio. *The Open: Man and Animals*. Stanford: Stanford University Press, 2004.

———. *The Use of Bodies*. Stanford: Stanford University Press, 2015.

Alberti, Leon Battista. *De re aedificatoria*. Florence: Accuratissime impressum opera Magistri Nicolai Laurentii Alamani, 1485.

Alfano Miglietti, Francesca. *Identità mutanti, dalla piega alla piaga*. Milano: Bruno Mondadori, 2008.

Anders, Günther. *Die Antiquiertheit des Menschen*. München: Verlag CH Beck, 1980.

Andersen, Karin. *Zoomanity*, edited by Betta Frigeri and Luca Panaro. Modena: Galleria Paggeriarte, 2006.

Andrea, Tortoreto. *Filosofia della fantascienza*. Sesto San Giovanni: Mimesis, 2018.

Andrieu, Bernard. *Le corps dispersé*. Paris: L'Harmattan, 1993.

Arecchi, Fortunato Tito. *Coerenza. Complessità. Creatività*. Roma: Di Renzo Editore, 2008.

Aristotle. "Categories." In *The Complete Works of Aristotle*, edited by Jonathan Barnes. Princeton: Princeton University Press, 2014.

Arthur, Brian W. *The Nature of Technology*. New York: Free Press, 2009

Augè, Marc. *Non-places. Introduction to an Anthropology of Supermodernity*. New York: Verso Books, 2008.

Augustine of Hippo. *The city of God*. Edinburg: T&T. Clark, 1913.

Austin, John L. *Sense and Sensibilia*. Oxford: Oxford University Press, Oxford 1962.

Bachelard, Gaston. *The Formation of the Scientific Mind*. Bolton-Manchester: Clinamen Press, 2002.

Bachimont, Bruno. "Between Formats and Data. When Comunications Becomes Recording." In *Towards a Philisophy of Digital Media*, edited by Alberto Romele and Enrico Terrone. London: Palgrave Macmillan, 2018.

Badiou, Alain. *Being and Event*. London: Continuum Intl Pub Group, 2006.

Balibar, Étienne. *The Philosophy of Marx*. New York: Verso Books, 1996.

Balibar, Étienne and Vittorio Morfino. *Il transindividuale*. Sesto San Giovanni: Mimesis 2014.
Barad, Karen. *Meeting the Universe Halfway*. Durham, NC: Duke University Press, 2007.
Barbaras, Renaud. *Introduction to a Phenomenology of Life*. Bloomington: Indiana University Press, 2022.
Barker, D. J. P. "Fetal Nutrition and Cardiovascular Disease in Later Life." *British Medical Bulletin* 53, 1 (January 1997): 96–108.
Barrau, Aurélien. *Le plus grand défi de l'histoire de l'Humanité*. Michel Lafon, Paris: 2020.
Bateson, Gregory. *Steps to an Ecology of Minds*. New York: Ballantine Books, 1976.
Battaglia, Luisella, ed. *Lo specchio oscuro. Gli animali nell'immaginario degli uomini*. Torino: Satyagraha, 1993.
Baudrillard, Jean. *The Consumer Society*. Thousand Oaks, CA: Sage Publications, 1998.
Bauman, Zygmunt. *Globalization: The Human Consequences*. New York: Columbia University Press, 1998.
———. *Decadence des intellectuels*. Nimes: J. Editions Jacqueline Chambon, 2007.
———. *Liquid Modernity*. Cambridge: Polity Press, 2000.
Baumgarten, Alexander G. *Theoretische Ästhetik*. Hamburg: F. Meiner, 1983.
Belk, Russell W. "Possessions and the Extended Self." *Journal of Consumer Research* 15, 1988.
Benasayag, Miguel. *El cerebro aumentado, el hombre disminuido*. Buenos Aires: Paidós, 2015.
Benasayag, Miguel and Gérard Schmit. *Les passions tristes*. Paris: Éditions La Découverte, 2003.
Benda, Julien. *The Betrayal of the Intellectuals*. Boston: Beacon Press, 1955.
Benjamin, Walter. *Il capitalismo come religione*. Sesto San Giovanni: Mimesis, 2014.
Benner, Susanne, and Gregor Lax, Paul J. Crutzen, Ulrich Pöschl, Jos Lelieveld, Hans Günter Brauch. *Paul J. Crutzen and the Anthropocene: A New Epoch in Earth's History*. New York: Springer, 2021.
Bergson, Henri. *Creative Evolution*. New York: Henry Holt and Company, 1911.
Berti, Enrico. *In principio era la meraviglia*. Bari: Laterza, 2007.
———. *Sumphilosophein*. Bari: Laterza, 2010.
Bocchi, Gianluca, and Mauro Ceruti. *La sfida della complessità*. Milano: Bruno Mondadori, 2007.
Boddice, Rob. *Anthropocentrism*. Boston: Brill, 2011.
Bodei, Remo. *Geometry of the Passions*. Toronto: University of Toronto Press, 2018.
Bolk, Louis. *Das Problem der Menschwerdung*. Jena: Fischer, 1926.
Borgato, Renata et al. *Nuovo Rinascimento. Idee per la formazione*. Milano: Franco Angeli, 2010.
Bourke, Joanna. *Fear: A Cultural History*. London: Virago Press, 2005.
Bove, Laurent. *La strategia del conatus*. Milano: Edizioni Ghibli, 2002.
Braidotti, Rosi. *Metamorphoses, Towards a Materialist Theoy of Becoming*. Cambridge: Polity, 2002.
———. *Per una politica affermativa. Itinerari etici*. Sesto San Giovanni: Mimesis, 2017.

———. *Posthuman Knowledge*. Cambridge: Polity, 2019.
———. *The Posthuman*. Cambridge: Polity, 2013.
Brian, Arthur W. *The Nature of Technology*. New York: Free Press, 2011.
Bruni, Marco. *La natura divisa. Hans Jonas e la questione del dualismo*. Roma: Inschibboleth, 2020.
Bruno, Giordano. *Cause, Principle, and Unity*. Cambridge: Cambridge University Press, 1998.
Buber, Martin. *Das dialogische Prinzip*. Heidelberg: Lambert Schneider, 1984.
Burke, Edmund. *A Philosophical Enquiry into the Origin of Our Ideas of the Sublime and Beautiful*. Chicago: Independently published, 2020.
Burkett, Paul. *Marx and Nature*. New York: St. Martin's Press, 1999.
Caffo, Leonardo. *The Contemporany Posthuman*. Cambridge: Ethics Press 2022.
Campbell, Donald T. "Evolutionary Epistemology." In *The Philosophy of Karl R. Popper*, edited by Paul A. Schilpp, 412–63. LaSalle IL: Open Court Publishing Company, 1974.
Campo, Enrico. *La testa altrove*. Roma: Donzelli, 2020.
Canguilhem, Georges. *The normal and the pathological*. New York: Zone Books, 1989.
Capucci, Pier Luigi. *Il corpo tecnologico*. Bologna: Baskerville, 1994.
Caronia, Antonio. *Il corpo virtuale*. Roma: Muzzio, 1996.
Carr, David, and Jan Steutel. *Virtue Ethics and Moral Education*. London: Routledge, 1999.
Carr, Nicholas. *The Shallows: What the Internet Is Doing to Our Brains*. New York: W.W. Norton, 2010.
Christen, Yves. *L'animal est-il une personne?* Paris: Flammarion, 2011.
Cimatti, Felice. *Filosofia dell'animalità*. Bari: Laterza, 2013.
Colborn, Theo, Dianne Dumanoski, and John Peterson Myers. *Our Stolen Future*. New York: Dutton, 1996.
Crain, Drew, and Louis J. Guillette Jr. "Reptiles as Models of Contaminant-induced Endocrine Disruption." *Animal Reproduction Science* 53 (1998): 77–86.
Cregan-Reid, Vybarr. *Primate Change: How the World We Made Is Remaking Us*. London: Cassell, 2018.
Damasio, Antonio R. *Looking for Spinoza*. Boston: Mariner Books, 2003.
———. *Descartes' Error: Emotion, Reason, and the Human Brain*. New York: HarperCollins, 1994.
Dawkins, Richard. *The Selfish Gene*. Oxford: Oxford University Press, 1976.
De Benoist, Alain. *Des animaux et des hommes*. Billière: Alexipharmaque, 2010.
De Gainza, Mariana. "Il limite e la parte: i confini dell'interiorità spinoziana." In *Il transindividuale*, edited by Balibar, Étienne and vittorio Morfino. Sesto San Giovanni: Mimesis 2014.
De Grey, Aubrey, and Michael Rae. *Ending Aging*. New York: St. Martin's Press, 2008
De Kerkhove, Derrick. *Brainframes: Technology, Mind and Business*. Utrecht: Bosch & Keuning, 2019.
Debord, Guy. *The Society of the Spectacle*. Cambridge: Unredacted Word, 2021.
Dehaene, Stanislas. *How We Learn*. New York: Viking, 2020.

Deitch, Jeffrey. *Post-Human*. Torino: Castello di Rivoli Museo d'Arte Contemporanea, 1992.
Del Noce, Augusto. *Il suicidio della rivoluzione*. Milano: Corriere della Sera, 2012.
Del Rey, Angélique. *La tyrannie de l'évaluation*. Paris: La Decouverte, 2013.
Del Val, Jaime. *Ecology of Affects: From Bioethic to Kinethics*. 4th Beyond Humanism Conference, Dubrovnik 2012
Deleuze, Gilles, and Félix Guattari. *A Thousand Plateaus: Capitalism and Schizophrenia*. Minneapolis: University of Minnesota Press, 1997.
———. *Cosa può un corpo? Lezioni su Spinoza*. Verona: Ombre Corte, 2010.
———. *Difference and Repetition*. London: Athlone Press, 1994.
———. *Foucault*. Minneapolis: University of Minnesota Press, 1988.
———. *Instincts & Institutions*. Paris: Hachette, 1955.
———. *Nietzsche and Philosophy*. New York: Columbia University Press, 2006.
———. *Spinoza: Practical Philosophy*. San Francisco: City Lights Books, 2001.
Demichelis, Lelio. *La grande alienazione*. Milano: Jaca Book, 2018.
Dennett, Daniel. 1995. *Darwin's Dangerous Idea: Evolution and the Meanings of Life*. New York: Simon & Schuster.
———. *Kinds of Minds: Toward an Understanding of Consciousness*. New York: Basic Books, 1996.
Derrida, Jacques. *Margins of philosophy*. Chicago: University of Chicago Press, 1982.
———. *Of Grammatology*. Baltimore: Johns Hopkins University Press, 1976.
———. *The Animals That Therefore I Am*. New York: Fordham University Press, 2008.
Descartes, René. *Discourse on Method*. New York: Dutton, 1951.
———. *Meditations metaphysiques*. Paris: Presses Universitaire de France, 1970.
Descola, Philippe. *Beyond Nature and Culture*. Chicago: University of Chicago Press, 2014.
———. *L'écologie des autres: L'anthropologie et la question de la nature*. Paris: Éditions Quae, 2016.
———. *Par-delà nature et culture*. Paris: Gallimard, 2005.
Deutsch, David. *The Fabric of Reality*. London: Penguin Books 1998
Diamond, Jared. *Collapse: How Societies Choose to Fail or Succeed*. London: Penguin Books, 2011.
———. *Guns, Germs and Steel: The Fates of Human Societies*. New York: W.W. Norton & Company, 1997.
Douglas, Angela E. "Symbiosis as a General Principle in Eukaryotic Evolution." *Cold Spring Harb Perspect Biol*. 6 (February 2014).
Douglas, Mary. *Purity and Danger*. New York: Routledge and Kegan Paul, 1966.
Dublin, Max. *Futurehype: The Tyranny of Prophecy*. Harmondsworh: Viking, 1989.
Dupré, John, and Maureen A. O'Malley. "Varieties of Living Things: Life at the Intersection of Lineages and Metabolism." *Philosophy, Theory, and Practice in Biology* 1, 3 (2009).
Dusi, Nicola, and Cosetta G. Saba, *Matthew Barney, polimorfismo, multimodalità, neobarocco*. Milano: Silvana Editoriale, 2012.

Eagleman, David. *Livewired: The Inside Story of the Ever-Changing Brain*. New York: Pantheon, 2020.
Easterling, Keller. *Medium Design: Knowing How to Work on the World*. New York: Verso Books, 2021.
Edelman, Gerald M. *Neural Darwinism*. New York: Basic Books, 1987.
Ehrlich, Paul R. *The Population Bomb*. New York: Ballantine Books, 1968.
Eibl-Eibesfeldt, Irenäus. *Human Ethology*. New York: Aldine De Gruyter, 1989.
———s. *Love and Hate: The Natural History of Behavior Patterns*. London: Routledge, 1997.
———. *The Biology of Peace and War: Men, Animals, Aggression*. New York: Viking Press, 1979.
Eliade, Mircea. *Shamanism: Archaic Techniques of Ecstasy*. Princeton: Princeton University Press, 1964.
Esposito, Roberto. *Communitas. Origine e destino della comunità*. Torino: Einaudi, 2006.
———. *Immunitas: The Protection and Negation of Life*. Cambridge: Polity, 2011.
Ferrai, Lucia. *La soggettività Cyborg*. Trento: Tangram Edizioni Scientifiche, 2013.
Ferrando, Francesca. *Il Postumanesimo Filosofico e le sue Alterità*. Pisa: Ets, 2016.
———. *Philosophical Posthumanism*. London: Bloomsbury Publishing, 2020.
Ferrante, Alessandro, and Jole Orsenigo. *Dialoghi sul postumano. Pedagogia, filosofia e scienza*. Sesto San Giovanni: Mimesis 2017.
Ferraris, Maurizio. "Esistere è resistere," in *Bentornata Realtà*, edited by Mario De Caro and Maurizio Ferraris. Torino: Einaudi, 2012.
———. L'esplosione della registrazione." In *Filosofia del digitale*, edited by Luca Taddio and Gabriele Giacomini. Sesto San Giovanni: Mimesis, 2020.
Fiedler, Leslie. *Freaks: Myths and Images of the Secret Self*. New York: Anchor, 1993.
Foer, Jonathan Safran. *We Are the Weather: Saving the Planet Begins at Breakfast*. New York: Macmillan, 2019.
Foucault, Michel. *Discipline and Punish*. New York: Random House, 1977.
———. *The Order of Things*. New York: Pantheon Books, 1970.
Francis, Gavin. *Adventures in Human Being*. London: Profile Books, 2015.
Fredkin, Edward. "Digital Mechanism: An Informational Process Base on Reversible Universal Cellular Automata." *Physica D: Nonlinear Phenomena* 45 (September 1990): 254–70.
Freud, Sigmund. *Mass Psychology: And Other Writings*. New York: Penguin, 2004.
Fukuyama, Francis. *Our Posthuman Future: Consequences of the Biotechnology Revolution*. New York: Farrar, Straus and Giroux, 2002.
Galimberti, Umberto. *Psiche e techne. L'uomo nell'età della tecnica*. Milano: Feltrinelli, 1999.
Garin, Eugenio. *L'umanesimo italiano*. Bari: Laterza, 1958.
Gavin, Francis. *Shapeshifters*. London: Profile Books, 2018.
Gawdat, Mo. *Scary Smart*. London: Bluebird Publishing, 2021.
Gehlen, Arnold. *Man: His Nature and Place in the World*. New York: Columbia University Press, 1988.
Geoffroy Saint-Hilaire, Etienne. *Cours de l'histoire naturelle des mammiféres*. Paris: Pichon et Didier, 1829.

Gibson, William. *Neuromancer*. New York: Ace Books, 1984.
Gilbert, Scott F. *Developmental Biology*. Sunderland: Sinauer Associates, 2000.
Girard, René. *The Scapegoat*. Baltimore: John Hopkins University Press, 1989.
Godfrey-Smith, Peter. *Theory and Reality. An Introduction to the Philosophy of Science*. Chicago: University of Chicago Press, 2021.
Goldberg, Sanford C. *The Brain in a Vat*. Cambridge: Cambridge University Press, 2016.
Gostin, Lawrence O. *Global Health Law*. Cambridge: Harvard University Press, 2014.
Gottschalk, Simon. "The Infantilization of Western Culture." *The Conversation* 2018. https://theconversation.com/the-infantilization-of-western-culture-99556.
Gould, Stephen Jay. *Ontogeny and Phylogeny*. Cambridge: Belknap Press of Harvard University Press, 1977.
———. *Wonderful Life*. New York: W.W. Norton & Company, 1990.
Gould, Stephen Jay, and Elisabeth Vrba. "Exaptation—A Missing Term in the Science of Form." *Paleobiology* 8, 1 (1982): 4–15.
Gould, Stephen Jay, and Richard C. Lewontin. "The Spandrels of San Marco and the Panglossian Paradigm: A Critique of the Adaptationist Programme." *Proceedings of the Royal Society B: Biological Sciences* 205, (1979): 581–98.
Greenfield, Susan. *Mind Change: How Digital Technologies Are Leaving Their Mark on Our Brains*. New York: Random House, 2015.
Griffin, Donald. *Animal Minds*. Chicago: University of Chicago Press, 2001.
Habermas, Jürgen. *Ach, Europa*. Berlin: Suhrkamp Verlag, 2008.
Hall, Edward T. *The Hidden Dimension*. New York: Anchor, 1990.
Hamilton, William D. "The Genetical Evolution of Social Behaviour." *Journal of Theoretical Biology* 7 (July 1964).
Han, Byung-Chul. *The Agony of Eros*. Cambridge: The MIT Press, 2017.
———. *The Expulsion of the Other: Society, Perception and Communication Today*. Cambridge: Polity, 2018.
———. *Non-things*. Cambridge: Polity, 2022.
Harari, Yuval N. *Sapiens: A Brief History of Humankind*. New York: Harper, 2015.
Haraway, Donna. *Staying with the Trouble. Making Kin in Chthulucene*. Durham: Duke University Press, 2016.
———. *A Cyborg Manifesto*. Minneapolis: University of Minnesota Press, 2016.
———. *When Species Meet*. Minneapolis: University of Minnesota Press, 2008.
Hardin, Garrett. "Lifeboat Ethics: The Case against Helping the Poor." *Psychology Today* 8 (September 1974): 38–43.
Hartmann, Nicolai. *New Ways of Ontology*. Westport: Praeger Publishers, 1942.
Hayles, Katherine N. *How We Became Posthuman*. Chicago: The University of Chicago Press, 1999.
Hegel, Georg W. F. *The Phenomenology of Spirit*. Notre Dame IN: University of Notre Dame, 2019.
Heidegger, Martin. *Being and Time*. New York: Harper, 1962.
———. *Nietzsche*. San Francisco: Harper One, 1991.
———. *Sein und Zeit*. Tubinga: M. Niemeyer, 1927.

———. *The Fundamental Concepts of Metaphysics*. Bloomington: Indiana University Press, 2001.
Held, Virginia. *The Ethics of Care*. Oxford: Oxford University Press, 2006.
Henry, Michel. *Material Phenomenology*. New York: Fordham University Press, 2008.
Herder, Johann Gottfried. *Ideen zur Geschichte der Menschheit*. Amsterdam: F.A. Brockhaus, 1869.
———. *Abhandlung über den Ursprung der Sprache*. München: C. Hanser, 1978.
Hobbes, Thomas. *Leviathan*. Oxford: James Thornton, 1881.
Hobsbawm, Eric. *The Age of Extremes: The Short Twentieth Century*. London: Micheal Joseph, 1994.
Hofstadter, Douglas. *Gödel, Escher, Bach: an Eternal Golden Braid*. New York: Basic Books, 1979.
Hughes, Howard C. *Sensory Exotica: A World Beyond Human Experience*. Cambridge: Bradford Books, 1999.
Hughes, James. *Citizen Cyborg*. Cambridge: Westview Press, 2004.
Hughes, Thomas P. *Human-Built World: How to Think about Technology and Culture*. Chicago: University of Chicago Press, 2005.
Hume, David. *A Treatise of Human Nature*. London: J.M. Dent & Son, 1956.
Husserl, Edmund. *Ideas: General Introduction to Pure Phenomenology*. Oxfordshire: Taylor & Francis, 2012.
Huxley, Aldous. *The Doors of Perception and Heaven and Hell*. New York: Harper Perennial, 2004.
Huxley, Julian. *New Bottles for New Wine*. London: Chatto & Windus, 1957.
Hyde, Lewis. *Trickster Makes This World: Mischief, Myth and Art*. New York: Farrar, Straus and Giroux, 2010.
Ingold, Tim. "Culture and the Perception of the Environment," in *Bush Base: Forest Farm*, edited by Elisabeth Croll and David Parkin, 39–56. London: Routledge, 1992.
Irwin, William. *The Matrix and Philosophy: Welcome to the Desert of the Real*. Chicago: Open Court Publishing Company, 2002.
Jacob, François. "Bricolage and the Possible." *Nottingham French Studies* 59, no. 3 (Winter 2020): 333–49.
Jaeggi, Rahel. *Alienation*. New York: Columbia University Press, 2014.
Jha, Prem Shankar. *Dawn of the Solar Age: An End to Global Warming and to Fear.* Thousand Oaks: Sage Publications, 1998.
Johnson, Steven. *Emergence: The Connected Lives of Ants, Brains, Cities and Software*. New York: Scribner, 2002.
Johnson-Laird, Philip N. *The Computer and the Mind. An Introduction to Cognitive Science*. London: William Collins Sons & Co, 1988.
Jonas, Hans. *The Imperative of Responsability*. Chicago: University of Chicago Press, 1985.
———. *The Phenomenon of Life: Towards a Philosophical Biology*. New York: Harper and Row, 1966.
Kant, Immanuel. *Critique of Practical Reason.* New York: Dover Publications, 2012.
———. *Critique of Pure Reason*. Chicago: Independently Published, 2021.

Kauffman, Stuart. *A World beyond Physics. The Emergence of Life*. Oxford: Oxford University Press, 2019.
Kelly, Kevin. *Out of Control*. New York: Basic Books, 1995.
———. *What Technology Wants*. New York: Penguin Books, 2011.
Kemp, Martin. *The Science of Art*. New Haven: Yale University Press, 1990.
Kirschner, Marc, and John Gerhart. "Evolvability." *Proc. Natl. Acad. Sci.* 95 (July, 1998): 8420–27.
Klein, Naomi. *This Changes Everything: Capitalism vs. the Climate*. London: Penguin Books, 2015.
Köhler, Wolfgang. *Gestalt Psychology*. New York: H. Liveright, 1929.
Kolbert, Elizabeth. *The Sixth Extinction: An Unnatural History*. London: Bloomsbury, 2015.
Kothari, Ashish, et al. *Pluriverso. Dizionario del post-sviluppo*. Salerno: Orthotes, 2021.
Kristeller, Paul Oskar. *The Philosophy of Marsilio Ficino*. New York: Colombia University Press, 1943.
Kropotkin, Pëtr. *Mutual Aid: A Factor of Evolution*. Bristol: Read & Co, 2020.
Kurzweil, Raymond. *The Singularity Is Near*. New York: Viking, 2005.
Lacan, Jacques. *Desire and Its Interpretation: The Seminar of Jacques Lacan*. Book VI. Cambridge: Polity, 2021.
Lanier, Jaron. *Who Owns the Future?* New York: Simon & Schuster, 2013.
Laporte, Jean. *Le rationalisme de Descartes*. Paris: Presses Universitaire de France, 1997.
Lasch, Christopher. *The Culture of Narcissism*. New York: W.W. Norton & Company, 1979.
Latour, Bruno. *Politics of Nature*. Cambridge: Harvard University Press, 2004.
Leakey, Richard. *The Origin of Humankind*. New York: BasicBooks, 1994.
Leary, Timothy, Ralph Metzner, and Richard Alpert. *The Psychedelic Experience*. New York: Citadel Press, 1964.
LeDoux, Joseph. *Synaptic Self*. New York: Viking, 2002.
Leibniz, Wilhelm G. *Discourse on Metaphysics*. Chicago: Open Court Publishing Company, 1980.
Lembke, Anna. *Dopamine Nation: Finding Balance in the Age of Indulgence*. Boston: Dutton, 2021.
Leonhard, Gerd. *Technology vs Humanity*. Berlin: Vahlen, 2016.
Leopardi, Giacomo. *Zibaldone*. Roma: Newton, 1997.
Lequyer, Jules. *Translation of Works of Jules Lequyer*. New York: Edwin Mellen, 1998.
Lévi-Strauss, Claude. *Totemism*. Boston: Beacon Press, 1963.
Levi-Montalcini, Rita. *Elogio dell'imperfezione*. Milano: Garzanti, 1987.
Levinas, Emmanuel. *Alterity and Transcendence*. New York: Columbia University Press, 2001.
———. *Discovering Existence with Husserl*. Evanston: Northwestern University Press, 1998.
———. *Entre Nous: On Thinking-of-the-Other*. New York: Columbia University Press, 2000.

Lévy, Pierre. *Les technologies de l'intelligence. L'avenir de la pensée a l'ére informatique*. Paris: Editions La Découverte, 1990.
Lewontin, Richard. "The Organism as the Subject and Object of Evolution." *Scientia* 118, (1983): 63–82.
Lichtenberg, Joseph D. *Psychoanalysis and Motivation*. Hillsdale: The Analytic Press, 1989.
Lindsay, Peter H., and Donald A. Norman. *Human Information Processing: An Introduction to Psychology.* New York: Academic Press, 2013.
Longo, Giuseppe O., and Andrea Vaccaro. *Bit Bang. La nascita della filosofia digitale*. Milano: Apogeo, 2013.
———. *Matematica e senso*. Sesto San Giovanni: Mimesis, 2021.
Lorenz, Konrad. *Behind The Mirror: A Search for a Natural History of Human Knowledge*. Boston: Houghton Mifflin, 1978.
———. *Die Rückseite des Spiegels*. München: Piper, 1973.
Macaulay Trevelyan, George. *English Social History*. London: Longmans, Green & Co., 1942.
Macrì, Teresa. *Il corpo postorganico*. Genova: Costa & Nolan, 1996.
Magdoff, Fred, and Brian Tokar. *Agriculture and Food in Crisis: Conflict, Resistance, and Renewal*. New York City: Monthly Review Press, 2010.
Malinowski, Bronislaw. *A Scientific Theory of Culture and Other Essays.* Chapel Hill: University of North Carolina Press, 1973.
Manetti, Giannozzo. *De dignitate et excellentia hominis*. Padova: Editrice Antenore. 1975.
Mann, Geoff and Joel Wainwright. *Climate Leviathan*. New York: Verso Books, 2020.
Marchesini, Roberto. "Knowledge and Different Levels of Reality." *Philosophical Reading* VI, 2 (Summer 2014): 53–64.
———. "Zoomimesis: How Birds Taught Us That We Can Fly." *Journal of Posthumanism Philosophy* 2, 2 (June 2022): 115–24.
———. *Beyond Anthropocentrism, Thoughts for a Post-Human Philosophy*. Udine: Mimesis International, 2018.
———. *Eco-ontologia. L'essere come relazione*. Bologna: Apeiron, 2018.
———. *Essere un corpo*. Modena: Stem Mucchi Editore, 2020.
———. *Estetica Postumanista*. Sesto San Giovanni: Meltemi, 2019.
———. *Le radici del desiderio*. Marchesini. Bologna: Apeiron, 2021.
———. *Oltre il Muro*. Roma: Muzzio, 1996.
———. *Over the Human: Post-humanism and the Concept of Animal Epiphany*. New York: Springer 2017.
———. *Post Human.* Torino: Bollati Boringhieri, 2002.
———. *Tecnosfera. Proiezioni per un futuro postumano*. Roma: Castelvecchi, 2017.
———. *Teriomorfismo*. Bologna: Apeiron, 2010.
———. *The Creative Animal*. New York: Palgrave Macmillan, 2022.
———. *The Virus Paradigm*. Cambridge: Cambridge University Press, 2021.
Margulis, Lynn. *Symbiosis as a Source of Evolutionary Innovation: Speciation and Morphogenesis.* Cambridge: MIT Press, 1991.
———. *Symbiosis in Cell Evolution*. San Francisco: W.H. Freeman, 1981

———. *Symbiotic Planet: A New Look at Evolution*. New York: Basic Books, 1999.
Marion, Jean-Luc and Emilie Tardivel. *Fenomenologa del dono*. Brescia: Morcelliana, 2018.
Maturana, Humberto and Francisco Varela. *Autopoiesis and Cognition: The Realization of the Living*. Dordrecht: Reidel Publishing Company, 1985.
Mayr, Ernst. *The Growth of Biological Thought*. Cambridge: Belknap Press, 1985.
McLuhan, Marshall. *Understanding Media: The Extensions of Man*. New York: McGraw Hill, 1964.
Merleau-Ponty, Maurice. *Nature*. Evanston: Northwestern University Press, 2003.
———. *Phénoménologie de la perception*. London: Routledge, 2013.
Midgley, Mary. *Heart and Mind: The Varieties of Moral Experience*. London: Routledge, 1981.
Milgram, Stanley. *Obedience to Authority*. New York: Harper & Row, 2003.
Monod, Jacques. *Chance and Necessity*. New York: Vintage Books, 1972.
Montanari, Massimo, and Bruno Andreolli. *Il Bosco nel Medioevo*. Padova: Clueb, 1988.
Moore, Jason W., ed., *Anthropocene or Capitalocene, Nature, History and the Crisis of Capitalism*. Okland: PM Press, 2016.
More, Max. "A Letter to Mother Nature." In *The Transhumanist Reader*, edited by Max More and Natasha Vita-More, 449–50. Hoboken NJ: John Wiley & Sons, 2013.
Morgan, Hugh. D. et al. "Epigenetic Inheritance at the Agouti Locus in the Mouse." *Nature Genet*. 23 (1999): 314–18.
Morin, Edgar. *L'Unité de l'homme: essais de discussion*. Paris: Seuil, 1974.
———. *La Tête bien faite. Pensere la réforme, reformer la pensée*. Paris: Seuil, 1999.
Mortari, Luigina. *Filosofia della cura*. Milano: Raffaello Cortina Editore, 2015.
Morton, Timothy. *Hyperobjects*. Minneapolis: University of Minnesota Press, 2013.
Mulgan, Geoff. *Big Mind*. Princeton: Princeton University Press, 2018.
Mumford, Lewis. *Technics and Civilization*. New York: Harcourt, 1934
———. *The Myth of the Machine*. New York: Harcourt, 1967.
Naess, Arne. *Ecology of Wisdom*. London: Penguin Books, 2016.
———. *Ecology, Community and Lifestyle*. Cambridge: Cambridge University Press 1991.
Nagel, Thomas. "What Is It Like to Be a Bat?," *Philosophical Review* 73 (April 1974): 435–50.
Nancy, Jean-Luc. *Being Singular Plural*. Stanford: Stanford University Press, 2000.
Negri, Carlo. *Deleuze e il postumano*. Napoli: Orthotes Editrice, 2020
Niessen, Bertram. *Sociale, digitale: Trasformazione della cultura e delle reti*. Milano: Doppiozero, 2013.
Nietzsche, Friedrich. *Beyond Good and Evil*. Charlotte: Information Age Publishing, 2020.
———. *The Gay Science*. New York: Dover Publication, 2020.
———. *The Birth of Tragedy*. New York: Dover, 1995.
———. *The Dionysian Vision of the World*. Minneapolis: University of Minnesota Press, 2012.
———. *The Will to Power*. Chicago: Independently Published, 2017.

———. *Thus Spoke Zarathustra*. New York, Modern Library, 1917.
Noble, David F. *The Religion of Technology. The Divinity of Man and the Spirit of Invention*. New York: Penguin Books, 1997.
Noble, Denis. *The Music of Life: Biology beyond Genes*. Oxford: Oxford University Press 2008.
Norris, Pippa. *Digital Divide: Civic Engagement, Information, Poverty, and the Internet Worldwide*. Cambridge: Cambridge University Press, 2001.
Nye, David E. *American Technological Sublime*. Cambridge: MIT Press, 1996.
O'Connell, Mark. *To Be a Machine*. New York: Anchor, 2018.
Odling-Smee, John F. "Niche Constructing Phenotypes." In *The Role of Behaviour in Evolution*, edited by Henry C. Plotkin. Cambridge: MIT Press, 1988.
Odling-Smee, F. John, Kevin N. Laland, and Marcus W. Feldman. "Niche construction." *Am. Nat.* 147 (2017): 641–48.
Odum, Eugen P., and Gary W. Barrett. *Fundamentals of Ecology*. Belmont, CA: Thomson Brooks-Cole, 2005.
Olien, D. Michael. *The Human Myth: An Introduction to Anthropology*. New York: Harper & Row, 1978.
Onishi, Bradley. "Information, Bodies, and Heidegger: Tracing Visions of the Posthuman." *Sophia* 50, (December 2010): 101–12.
Ortega y Gasset, José. *Origen y epilogo de la filosofia*. Madrid: Alianza Editorial, 2007.
Oyama, Susan. *The Ontogeny of Information*. Durham: Duke University Press, 2000.
Packard, Vance. *The Hidden Persuaders*. New York: D. Mckay Co, 1957.
Pallavicini, Federica. *Psicologia della realtà virtuale*. Milano: Mondadori, 2020.
Passmore, John. *Man's Responsibility for Nature*. London: Duckworth, 1974.
Patel, Raj. *The Value of Nothing*. London: Portobello Books, 2009.
Patocka, Jan. *Le monde naturelle comme problème philosophique*. Paris: Vrin, 2016.
Pepperell, Robert. "The Posthuman Manifesto." *Kritikos* 2, (February 2005).
Piaget, Jean. *Psychologie et epistemologie*. Paris: Gonthier, 1970.
Pico della Mirandola, Giovanni. *Oratio de homini dignitate*. Firenze: Vallecchi, 1942.
Pietrzykowski, Tomasz. *Personhood Beyond Humanism*. New York: Springer, 2018.
Pigliucci, Massimo, and Gerd B. Müller. *Evolution: The Extended Synthesis*. Cambridge: MIT Press, 2010.
Pinker, Steven. *The Language Instinct: How the Mind Creates Language*. Chicago: Unabridged, 2014
Pinto Minerva, Franca, and Rosa Gallelli. *Pedagogia e post-umano. Ibridazioni identitarie e frontiere del possibile*. Roma: Carocci 2004.
Plessner, Helmuth. *Levels of Organic Life and the Human*. New York: Fordham University Press, 2019.
Polkinghorne, John. *One World*. Princeton: Princeton University Press, 1987.
Popitz, Heinrich. *Phenomena of Power*. New York: Columbia University Press, 2017.
Popper, Karl. *Conjectures and Refutations: The Growth of Scientific Knowledge*. London: Hassell Street Press, 2021.
———. *The Logic of Scientific Discovery*. Routledge Classic 2002
———. *The Open Society and Its Enemies*. New York: Harper, 1962.
Porges, Stephen W. *The Polyvagal Theory*. New York: W.W. Norton & Company, 2011.

Portmann, Adolf. "Die Ontogenese des Menschen als Problem der Evolutionsforschung." *Verhandlung der Schweizer Naturforschenden Gesell schaft*, (1936): 224–41.

Prigogine, Ilya, and Gregoire Nicolis. *Self-Organization in Nonequilibrium Systems: From Dissipative Structures to Order through Fluctuations*. Hoboken, NJ: Wiley-Blackwell, 1977.

Prigogine, Ilya, and Isabelle Stengers. *La Nouvelle alliance, Métamorphose de la science*. Paris: Édition Gallimard, Isabelle 1979.

Prodomo, Raffaele. *25 anni di bioetica a Napoli*. Sesto San Giovanni: Mimesis 2020.

Pulcini, Elena. *Care of the World*. New York: Springer, 2012.

Putnam, Hilary. *The Threefold Cord: Mind, Body and World*. New York: Columbia University Press, 1999.

Quammen, David. *Spillover: Animal Infections and the Next Human Pandemic*. New York: W.W Norton & Company, 2012.

———. *The Tangled Tree*. New York: Simon & Schuster, 2018.

Randers, Jørgen. *2052: A Global Forecast for the Next Forty Years*. Hartford: Chelsea Green Publishing, 2012.

Regan, Tom. *The Case for Animal Rights*. Berkeley: University of California Press, 2004.

Remotti, Francesco. *Contro l'identità*. Bari: Laterza, 1996.

Rifkin, Jeremy. *Beyond Beef: The Rise and Fall of the Cattle Culture*. New York: Plume, 1993.

Rignani, Orsola. *Emergenze Post-umaniste dell'umano*. Sesto San Giovanni: Mimesis, 2014.

Rilke, Rainer Maria. *Duino Elegies*. Manchester: Carcanet Press, 1993.

Ritzer, George. *The McDonaldization of Society*. Newbury Park: Sage Publications, 2014.

Romanes, George. *Animal Intelligence*. New York: Appleton and Company, 1892.

Rorty, Richard. *Philosophy and the Mirror of Nature*. Princeton: Princeton University Press, 2017.

Rosa, Hartmut. *Alienation and Acceleration*. Aarhus: Aarhus University Press, 2010.

Rosati, Gianpiero. *Narciso e Pigmalione. Illusione e spettacolo nelle Metamorfosi di Ovidio*. Pisa: Scuola Normale Superiore, 2017

Ross, Sheldon M. *Probability Models for Computer Science*. New York: Academic Press, 2001.

Rucker, Rudy. "Transrealist Manifesto." *The Bulletin of the Science Fiction Writers of America*, 82, 1983.

Ruesh, Hans. *Imperatrice nuda*. Milano: Rizzoli 1975.

Rullani, Enzo and Alberto F. De Toni, *Uomini 4.0: Ritorno al futuro. Creare valore esplorando la complessità*. Milano: Franco Angeli 2019.

Sagan, Carl, and Ann Druyan. *Pale Blue Dot: A Vision of the Human Future in Space*. New York: Ballantine Books, 1997.

Sandberg, Anders. "Morphological Freedom. Why We Not Only Want It, But Need It." In *The Transhumanist Reader*, edited by Max More and Natasha Vita-More. Hoboken NJ: Wiley-Blackwell, 2013.

Sapolsky, Robert. *Why Zebras Don't Get Ulcers*. New York: Henry Holt, 1994.

Sartre, Jean-Paul. *Sketch for a Theory of the Emotions*. London: Routledge, 2001.

Savulescu, Julian and Nick Bostrom. *Human Enhancement*. Oxford: Oxford University Press, 2009.

Sax, Boria. *Avian Illuminations*. London: Reaktion Books, 2021.

Scheler, Max. *Man's Place in Nature*. Boston: Beacon Press, 1961.

———. *The Human Place in the Cosmos*. Evanston: Northwestern University Press, 2009.

———. *The Nature of Sympathy*. Oxfordshire: Routledge & K. Paul, 1970.

Schopenhauer, Arthur. *The World as Will and Representation*. Cambridge: Cambridge University Press, 2010.

Schrödinger, Erwin. *What Is Life?* Cambridge: Cambridge University Press, 1940.

Scott, Gilbert, and David Epel. *Ecological Developmental Biology: The Environmental Regulation of Development, Health and Evolution*. Oxford: Oxford University Press, 2015.

Servier, Jean. *Histoire de l'utopie*. Paris: Gallimard, 1967.

Shannon, Claude Elwood. *The Mathematical Theory of Communication*. Urbana: University of Illinois Press, 1998.

Shaw, Debra Benita. *Posthuman Urbanism*. Lanham, MD: Rowman & Littlefield, 2017.

Shepard, Paul. *Nature and Madness*. Athens: University of Georgia Press, 1998.

Shimasaki, Craig. *Biotechnology Entrepreneurship*. Amsterdam: Elsevier, 2014.

Shipman, Pat. *The Invaders: How Humans and Their Dogs Drove Neanderthals to Extinction*. Cambridge: Belknap Press, 2017.

Shubin, Neil. *Your Inner Fish*. New York: Pantheon Books, 2009.

Shusterman, Richard. *Body Consciousness*. Cambridge: University Press, 2012.

Simmel, Georg. "The Metropolis and Mental Life." In *On Individuality and Social Forms: Selected Writings*, edited by Donald N. Levine, 324–39. Chicago: University of Chicago Press, 1971.

———. *Essay on Art and Aesthetics*. Chicago: University of Chicago Press, 2020.

———. *The Problem of the Philosophy of History*. New York: Free Press, 1977.

Simondon, Gilbert. *Individuation in Light of Notions of Form and Information*. Minneapolis: University of Minnesota Press, 2020.

———. *Sur la technique*. Paris: PUF, 2014.

Singer, Peter. *Animal Liberation*. New York: New York Review/Random House, 1975.

———. *The Animal Liberation Movement: Its Philosophy, Its Achievements and Its Future*. New York: Harper Collins, 1975.

Sisto, Davide. *Ricordati di me*. Torino: Bollati Boringhieri, 2020.

Sloterdijk, Peter. *Bubbles. Spheres Volume I: Microspereology*. Los Angeles: Semiotext(e), 2011.

———. *You Must Change Your Life*. Cambridge: Polity Press, 2013.

Snaza, Nathan. *Posthumanism and Educational Research*. Oxforshire: Taylor & Francis, 2016.

Snow, Charles Percy. *The Two Cultures and the Scientific Revolution*. Cambridge: Cambridge University Press. 1959.

Sorgner, Stefan L. *We Have Always Been Cyborg*. Bristol: Bristol University Press, 2023.

Spengler, Oswald. *Man and Technics*. New York: Knopf, 1931.
———. *The Decline of the West. Form and Actuality*. Budapest: Arktos Media, 2021.
Spinoza, Baruch. *Ethics*. New York: Dover, 1951.
Spitzer, Manfred. *Digitale Demenz*. München: Droemersche Th. Knaur Nachf, 2012.
Stein, Edith. *On the Problem of Empathy*. Leiden: Martinus Nijhoff, 1964.
Sterelny, Kim. *Dawkins vs. Gould: Survival of the Fittest*. Cambridge U.K.: ICON, 2007
Stern, Nicholas. *A Blueprint for a Safer Planet: How to Manage Climate Change and Create a New Era of Progress and Prosperity*. New York: Penguin Random House, 2009.
Stiegler, Bernard. *De la misère symbolique*. Paris: Flammarion, 2013.
———. *La Technique et le Temps. La faute d'Epiméthée*. Paris: Galilée, 1994.
———. *The Reenchantment of the World: The Value of Spirit Against Industrial Populism*. New York: Continuum, 2013.
Swan, Shanna H., and Stacey Colino. *Count Down*. New York: Scribner, 2020.
Taylor, Charles. *The Ethics of Authenticity*. Cambridge: Harvard University Press, 1991.
———. *The Malaise of Modernity*. Concord: Anansi, 1991.
Tenenti, Alberto. *Il senso della vita e l'amore della vita nel Rinascimento*. Torino: Einaudi, 1989.
Terrosi, Roberto. *La filosofia del postumano*. Genova: Costa & Nolan, 1997.
Thomas, Keith. *Man and the Natural World*. London: Penguin Books, 1983.
Thompson Klein, Julie. *Interdisciplinarity, History, Theory & Practice*. Detroit: Wayne State University Press, 1990.
Tinbergen, Niko. "On Aims and Methods of Ethology." *Zeitschrift für Tierpsychologie* 20 (March 1963): 410–33.
———. *Animal Behavior*. New York: Time Inc., 1965.
Tort, Patrick. *Darwin et la philosophie*. Paris: Kimé, 2004
Turkle, Sherry. *Alone Together*. New York: Basic Books, 2017.
Tymieniecka, Anna Teresa. *Phenomenology of Life: From the Animal Soul to the Human Mind*, Book II. *Analecta Husserliana*. New York: Springer, 2007.
Valentini, Nerino. *Mostruosità*. Mantova: Sometti Editoriale, 2016.
Valentini, Tommaso. "Genealogie del moderno e secolarizzazione" in *La filosofia italiana del Novecento*, edited by Onorato Grassi and Massimo Marassi. Sesto San Giovanni: Mimesis, 2015.
Van Gennep, Arnold. *The Rites of Passage*. Chicago: University of Chicago Press, 2019.
Van Potter, Rensselaer. *Bioethics: Bridge to the Future*. Englewood Cliffs, NJ: Prentice Hall, 1971.
Verducci, Daniela. "Education and the Ontopoietic Conception of Life." *Analecta Husserliana* 95, New York: Springer, 2008.
———. *La fenomenologia della vita di Anna-Teresa Tymieniecka*. Roma: Aracne editrice, 2012.
Vergine, Lea, and Giorgio Verzetti eds. *Il bello e le bestie: Metamorfosi, artifici e ibridi dal mito all'immaginario scientifico*. Milano: Skira Editore, 2004.

Vinge, Vernor. "The Coming Technological Singularity: How to Survive in the Post-Human Era." *Whole Earth Review* (Winter 1993).

Virilio, Paul. *Speed and Politics: An Essay on Dromology*. New York: Columbia University Press, 1986.

Viveiros de Castro, Eduardo. *Métaphysiques cannibales: Lignes d'anthropologie post-structurale*. Paris: Presses Universitaires de France, 2009.

Von Bayer, Hans C. *Information: The New Language of Science*. Cambridge: Harvard University Press 2004.

Von Bertalanffy, Ludwig. *General System Theory*. New York: George Braziller, 2015.

Von Goethe, Johann Wolfgang. *Faust*. New York: Modern Library, 1912

Von Uexküll, Jakob. *Streifzüge durch die Umwelten von Tieren und Menschen: Ein Bilderbuch unsichtbarer Welten. Bedeutungslehre*. Hamburg: Rowohlt, 1956.

Waddington, Conrad H. "Evolutionary adaptation." In *Evolution after Darwin*, edited S. Tax, vol. 1, 381–402. Chicago: University Chicago Press, 1959.

Waldrop, Mitchell. *Complexity: The Emerging Science at the Edge of Order and Chaos*. New York: Simon & Schuster, 1992.

Wallace, Alfred R. *Contributions to the Theory of Natural Selection*. London: Macmillan & Co, 1870.

Weisman, Alan. *The World Without Us*. London: Picador, 2008.

West-Eberhard, Mary Jane. *Developmental Plasticity and Evolution*. Oxford: Oxford University Press, 2003.

Wheeler, John A., and Kenneth Ford. *Geons, Black Holes & Quantum Foam*. New York: Norton & Company, 1998.

White Jr., Lynn. "The Historical Roots of Our Ecologic Crisis." *Science* 155, 3767 (March 1967): 1203–7.

Wilson, Edward O. *Sociobiology: The New Synthesis*. Cambridge: Harvard University Press, 1975.

Wittgenstein, Ludwig. *Philosophische Untersuchungen*. Oxford: Blackwell, 1953.

Wolfe, Cary. *Art and Posthumanism*. Minneapolis: University of Minnesota Press, 2022.

———. *What Is Posthumanism*? Minneapolis: University of Minnesota Press, 2009

Wolfram, Stephen. *A New Kind of Science*. Campaign, IL: Wolfram Media Inc, 2002.

Wolpert, Lewis. *The Unnatural Nature of Science*. Cambridge: Harvard University Press, 2000.

Zaltieri, Cristina. *L'invenzione del corpo*. Mantova: Negretto, 2009.

Žižek, Slavoj. *Welcome to the Desert of the Real*. New York: Verso, 2002.

# Index

# About the Author

Roberto Marchesini is director of the School of Human-Animal Interactions and the Center for the Study of Posthumanist Philosophy, both based in Bologna, Italy. He collaborates with the World Phenomenology Institute in Boston and has published his essays in the magazine *Analecta Husserliana*. Among his main publications: *Over the Human: Post-humanism and the Concept of Animal Epiphany* (Springer, 2017), *The Philosophical Ethology of Roberto Marchesini*, collected essays edited by Jeffrey Bussolini, Brett Buchanan, and Matthew Chrulew, (Routledge, 2017), *Dialogo Ergo Sum: My Pathway into Posthumanities* (University of Virginia Press, 2018), *Beyond Anthropocentrism* (Mimesis International, 2018), "Philosophical Ethology and Animal Subjectivity," in *Animality in Contemporary Italian Philosophy,* edited by Felice Cimatti and Carlo Salzani (Palgrave Macmillan, 2020), *The Virus Paradigm* (Cambridge University Press, 2021), *Critical Ethology and Post-Anthropocentric Ethics* (with Marco Celentano, Springer, 2021), and *The Creative Animal: How Every Animal Builds Its Own Existence* (Palgrave Macmillan, 2022).

www.ingramcontent.com/pod-product-compliance
Ingram Content Group UK Ltd.
Pitfield, Milton Keynes, MK11 3LW, UK
UKHW041838031026
15587UKWH00002B/845

9 781666 928228